Enhancing Repeatability and Reliability in High-Power Laser Beam Welding

Molina

Contents

Chapter 1

Introduction

This licentiate book addresses an aspect of production technology that is metal processing through fusion. A challenge to take up in production technology is to develop sustainable manufacturing processes to save both material and energy consumption. A metal fusion heat source with high efficiency is laser beam. The industrial demand to increase the productivity and improve the robustness of high-power laser beam welding (LBW) and laser beam additive manufacturing (LBAM) processes is driving huge research efforts towards the repeatability and reliability of the processes. Application of high-power continuous-wave laser beam for metal fusion in areas such as welding and additive manufacturing (AM) has increased during the last years with rapid pace. Continuous-wave lasers have proven their great performance in terms of their stability and power output during processing. However laser metal fusion is very complicated, and more detailed physical understanding would improve this process. The aim of this licentiate book is to reach better model reliability and deeper understanding of continuous-wave LBW of a Titanium alloy (Ti-6Al-4V) in conduction mode through a Computational Fluid Dynamics (CFD) approach.

1.1 Motivation

LBW is, in general, a difficult process to control due to the complex physical phenomena involved. In this process, several process parameters need to be specified. Among these parameters, one is attracting high attention in process development as it allows tuning the melt pool geometry through setting the energy input distribution; this is the beam shape profile. To better understand the effect of process parameters on the physical phenomena taking place along the metal fusion process in LBW and their impact on final parts geometry and quality, a self-consistent and verified modelling framework is needed. Here self-consistent means that the leading order physics should be predicted by the model rather than mimicked through adjustable parameters. And model verification implies confronting to experimental measurements to test the model predictability, e.g. in three space-dimensions and not only two if the problem cannot be reduced to a two-dimensional problem. Such a model would constitute a solid ground for providing a comprehensive understanding of the fundamentals of LBW. Not only does such understanding have academic motivations with respect to creating profound knowledge in the field, but also it has several important industrial motivations. The produced knowledge has important impact on industries that particularly deal with sensitive and expensive materials like GKN Aerospace, Alfa Laval, and Brogren Industries in Sweden. It could support the development of process control to enhance process repeatability and reliability. It would therefore help increasing productivity and reducing the cost of production through better understanding of the LBW which provides a proper set of process parameters and consequently leads to an improved quality of the product.

1.2 Research gaps

In this book the approach used is computational fluid dynamics. As the modelling and experimental approaches go hand in hand and supplement each other, collaborative work with experimentalists was also conducted. Three knowl-edge gaps were identified by analyzing the state-of-the-art through literature, as explained in more details hereafter:

(1) The effect of the beam shape on the process thermal flow is still ignored. The main reason is that the studies addressing this process parameter used to be experimental. Therefore the observations are restricted to the workpiece surface and to the post-solidified bead. The thermal flow within the workpiece bulk could not be accessed in these experimental studies. Thus, in addition to the need for process knowledge, evidence is also needed to confirm or infirm presumptions on the effect of the beam shape on the thermal flow made in the earlier experimental studies.

(2) There is lack of data for model validation. The standard validation approach is qualitative, and suited to a two-dimensional problem while the welding process is typically three-dimensional. A collaborative work with experimentalists was thus conducted to obtain a wider type of validation data to test the model. The comparison between calculation and experimental results did reveal model weaknesses. As a result the self-consistency of the CFD model with free surface tracking considered as

state of the art in the conduction mode LBW community needed to be questioned. It led to the identification of the next research gap.

(3) The modelling of the absorption of the laser beam energy by the material is over-simplified. It is reduced to a constant when modelling conduction mode LBW. This constant is set through a trial-and-error process in order to get the numerical results fitted with the experimental data for the process conditions studied. Sometimes, it is also re-used from one study to another with same metal alloy even though the process conditions can differ. Besides, previous studies conducted in physics to investigate the optical properties of metals have shown that the laser beam absorptance is not a constant value. For instance, it varies with the local surface conditions. There is therefore a need for improving the CFD model for conduction mode LBW, through predicting the absorptance.

The main focus of this book is therefore to investigate the two following aspects regarding LBW: shaped beam profile and laser absorptivity coefficient.

1.3 Objectives and research questions

Based on the industrial motivation and the identified research gaps discussed further in the next Chapter 3, the main objective of this research work is to better understand the melt pool thermo-hydrodynamic behaviour during LBW in conduction mode with a CFD approach. To achieve the research objective, the following research questions (RQ) are formulated and addressed:

(1) **RQ1**: How do different power density distributions in the laser beam spot modify the process physics during LBW in conduction mode?

(2) **RQ2**: How can the absorptivity of laser beam energy by a metal alloy be predicted when simulating LBW in conduction mode with CFD?

(3) **RQ3**: How does a predicted absorptance affect the process physics during LBW compared to a constant absorptivity coefficient?

Through addressing these research questions, knowledge that can support process development and process control can be provided. Furthermore, solid foundation for building upon with metal transfer to extend the modelling work to AM with laser and wire will be laid.

1.4 Methodology

In order to address the **RQ1**, the effect of different laser beam profiles (top-hat and Gaussian elliptical) on the flow and temperature fields inside the melt pool is first investigated. Then the variations in melt pool geometry with respect to each beam profile are studied. Regarding the **RQ2**, a mathematical method is presented to calculate the laser absorptivity. It includes the specification of the underlying assumptions and simplifications. The proposed method is then applied, and validated through confrontation to novel experimental measurements which include quantitative data characterizing of the melt pool free surface geometry. Comparisons are made between the results obtained from variable and constant laser absorptivity coefficients in order to address **RQ3**. The above-mentioned research questions will be expanded into several sub-questions in Chapter 2 based on the state of art knowledge.

To carry out this study, a solver is developed in an open-source software (Open-FOAM) for the simulation of welding of Ti-6Al-4V with laser. The developments have been added to the original built-in solver *interfoam*. The solver development procedure has been divided into the following steps:

- Implementing proper functions to calculate the material properties with respect to temperature.

- Implementing different laser power density distributions effective on the workpiece.

- Implementing a calculation process to better model the laser absorptivity.

- Making the model run with adaptive mesh refinement method (to decrease the computational cost).

It is also worth mentioning that in order to assess the reliability of the implementation at each step, experimental data are used.

1.5 Limitations and assumptions

There are some limiting factors in this work which originate either from the project requirements or from physical issues. The material is Ti-6Al-4V, which is the project requirement. Ti-6Al-4V is a Titanium alloy with a high strength-to-weight ratio, excellent corrosion resistance, and has wide application in aerospace industry. It has also poor thermal conductivity and is highly reactive to oxygen. Its thermo-physical properties are relatively well-documented compared to many other alloys. This research work does not go through the micro-structure analysis, however, the numerical results will provide the temperature distribution and melt pool geometry which could be used in micro-structure analysis. In order to build the model the following assumptions have been made:

With respect to the developed model:

- The flow and temperature fields are assumed to be laminar. Although there are some research works on the effect of turbulence on the melt pool thermo-hydrodynamic behaviour, the effective occurrence of this flow regime is not yet well-proved or well-documented.

- The effect of material vaporization is neglected, since the welding is in conduction mode, despite the fact that in a real application there could be small amount of metal vaporization (checked to be negligible in this study).

- The effect of surface oxidation is also ignored. The oxidation is kept as low as possible by providing sufficient amount of argon gas into the welding chamber.

With respect to the experimental test cases:

- Although in the experiments a flow of argon gas is blown into the welding chamber in order to maintain the concentration of argon gas close to 100%, such flow is ignored in the numerical simulations due to lack of information about the direction and magnitude of the flow. Instead natural convection is used with a reasonable heat transfer coefficient at the metal-gas interface.

- The laser wavelength is fixed 1060 nm for all cases.

- Metal transfer is not included in this licentiate book.

1.6 Licentiate overview

This licentiate book is organized as follows. In Chapter 1 an introduction about the project, research objectives, and methodology was provided. In Chapter 2 the LBW and the driving forces in conduction mode LBW are discussed. Chapter 3 provides complementary information about the state of the art discussed in the appended papers. The numerical method as well as the transport properties of the metal alloy are discussed in Chapter 4. In Chapter 5, different test cases are presented in order to show the solver development progress as well as the performance of the new implementations to the base solver. The summary of publications are presented in Chapter 6, and finally in Chapter 7 the research outcomes are summarized. The produced research papers are also appended at the end of this book.

Chapter 2

Laser beam welding

Laser is an efficient and widely-used heat source to fuse metals in welding. It has some great features such as: precise working with localised placing of the energy spot, ability of handling complicated joint geometry, minimal heat-affected zone, and low post-process operation times [1]. Laser welding is a common method for joining materials in many industries, such as aerospace and automotive. It involves complicated physics with different phenomena. These phenomena include the interaction of an electromagnetic wave with the metal alloy, material melting, re-solidification, and vaporization if there is high enough beam power density (laser power divided by its spot area), convection and radiation heat transfer, surface tension and Marangoni forces, and recoil pressure effect in case of material vaporization [1]. Understanding these features may help to improve the welding process and consequently may result in improved material properties. CFD is a powerful tool that can provide complementary process information within the metal bulk that would be difficult to access through experimental observation. This chapter gives a short overview of the fundamental aspects for setting the framework of this study regarding welding mode and the driving forces in LBW.

2.1 Welding modes

There are two distinct modes used in LBW: conduction mode and keyhole mode. The difference between these two modes basically arises from the amount of laser beam power density used and its potential to induce metal vaporization. The keyhole mode welding, which is characterized by intense vaporization, has some advantages such as large penetration depth, small heat affected zone (HAZ), and high production rate [2,3]. Despite these advantages, keyhole mode welding suffers from some critical drawbacks like high level of porosity, possible instability, and spatter [4]. The major reason for such disadvantages is the vaporization of the working material during welding due to the high laser power density [1]. On the other hand, conduction mode welding provides almost a spatter-free weld with little porosity and little cracks. It also creates a more stable welding process at which the laser heat source is more controllable compared to keyhole mode welding. Such capability can be applied to improve the weld quality and material micro-structure after solidification [5,6]. Two major drawbacks of the conduction mode welding are its low productivity and low heat absorption compared to keyhole mode welding. Table 2.1 presents a summary of advantages and drawbacks of LBW in conduction and keyhole modes [1].

Table 2.1: Advantages and drawbacks of LBW in conduction and keyhole modes [1].

Welding mode	Advantages	Drawbacks
Keyhole	High energy absorption High production rate Deep penetration welds	Less stable process High porosity High level of spattering
Conduction	Little vaporization Little crack and porosity Stable welding with controllable heat input Works fine with low-quality beam	Low production rate Low energy absorption High distortion of weld at high-power laser beam

Although many research works have been done on both conduction mode welding and keyhole mode welding both numerically and experimentally, the studies on the transition between these two modes are very limited. The conduction mode welding has two main characteristics compared to the keyhole mode welding. First, the vaporization is negligible during the process, and second the ratio of weld width to depth is lower than 0.5 [7,8]. Nevertheless, these definitions ignore the effect of welding speed, beam diameter, and laser beam profile. It has been claimed that when the laser power density is lower than 10^6 W/cm^2 the welding process is in conduction mode [9, 10], however this is valid only for specific laser beam wavelength and material and should be extrapolated when changing laser beam wavelength or material. Assuncao et al. [11] experimentally studied the effect of welding process parameters such as beam diameter, power density, and interaction time (see Eq.(3.3)) on the transition between conduction mode and keyhole mode (see Section 3.1). They used S355 mild steel as the working material and a fiber laser (with the wavelength of 1070 nm, a maximum power of 20 kW, and different beam diameters) as the heat source. They found that when increasing the beam power density there was no sharp transition between the conduction mode and the keyhole mode (see Figure 2.1). Furthermore, their results showed that welds in the transition region had features from both conduction and keyhole mode welding. They also found that there was no unique value for the weld width to depth ratio (aspect ratio) that could be considered as the transition value between the two different welding modes since the aspect ratio was dependent on the interaction time. Suder and Williams [12] showed that the two following parameters: interaction time and power density were not sufficient to characterize the LBW process in both conduction and keyhole modes. They thus defined a new parameter called specific point energy (product of power density, interaction time, and laser beam spot area, see Section 3.1). These authors used S355 low carbon steel as the working material and an IPG YLR-8000 CW fiber laser with maximum power of 8 kW. Their results revealed that the weld depth was mainly controlled by the power density and the specific point energy, whereas the weld width was controlled by the interaction time.

Cho et al. [13] numerically investigated the temperature and flow fields variations during laser spot welding in the transition mode. Their simulation results showed that the vaporization had a dominant effect on the weld pool surface temperature and flow pattern. They claimed that since there was a time lag between the net surface heat flux and the weld pool surface temperature, the weld pool flow continuously oscillated between upwards and downwards at the weld pool centre as the simulation time progressed. They also found that the orders of magnitude of the surface tension and recoil pressure were the same. The recoil pressure played an important role in the pool centre flow direction variation from a downward flow to an upward flow in case of concave weld pool surface.

In the previous research works, different authors used different process parameters, related to the laser beam, for categorizing the welding mode. Colegrove et al. [14] categorized the welding modes into three modes: conduction, mixed and keyhole (see Figure 2.2). Their results showed that in the conduction mode the vaporization was almost negligible and the melt pool flow pattern was re-

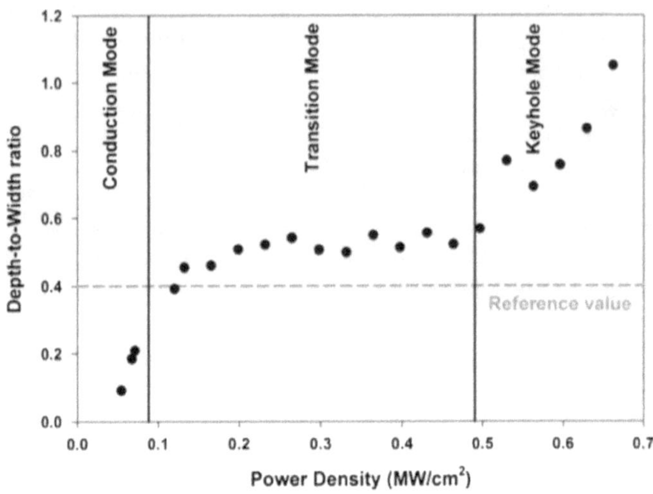

Figure 2.1: Variation of melt pool depth to width ratio with the power density for a beam diameter of 1.18 mm and an interaction time of 10 ms [11].

sponsible for the weld shape, whereas in the mixed regime vaporization started playing a role in melt pool shape, however the effect was not very significant.

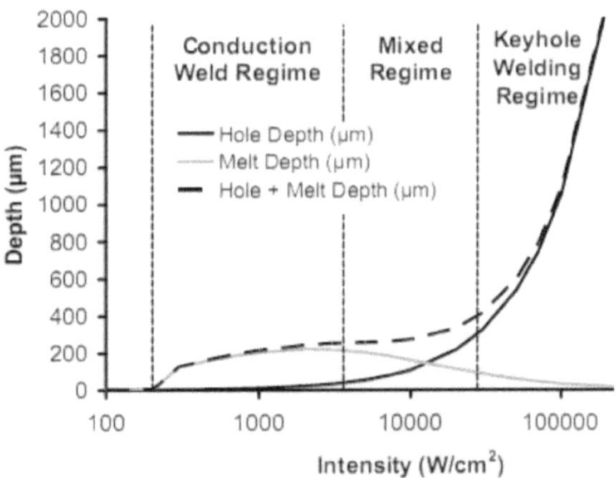

Figure 2.2: Weld depth as a function of the power density for different welding modes [14].

Ayoola et al. [15] proved that the weld geometry could be controlled by the spatial and temporal distribution of the laser energy on the surface of the work-piece. In other words, a desired bead size could be achieved by changing the laser beam specifications (e.g. diameter, profile, power) through properly tuning the optical set-up. These authors also found that in the conduction mode the bead geometry was almost independent of the beam diameter (in the specific range of process parameters studied). They observed that at a specific energy density the transition mode began and as a result the vaporization governed the process. It has also been shown that the depth of penetration in pure conduction mode welding was controlled by the power density and interaction time and was independent of the circular beam diameter, whereas the effect of beam diameter became more important when the transition mode began (see Figure 2.3). This meant that at different welding modes different parameters were more important and could be used to control the bead geometry.

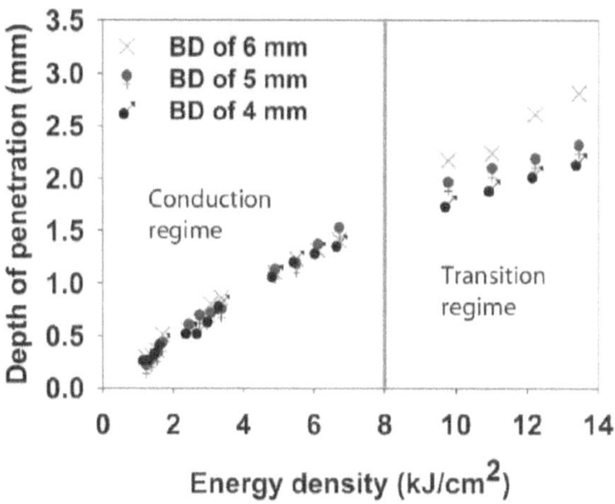

Figure 2.3: Effect of the energy density (laser beam power density times the interaction time) on the welding mode at different circular beam diameters [15].

In this book the emphasis is on conduction mode LBW. Therefore the vaporization and evaporation of metal during the process are neglected.

2.2 Driving forces in conduction mode laser beam welding

There are various driving forces that affect the thermo-hydrodynamic behaviour of the melt pool. There are several forces in conduction mode welding such as buoyancy, gravity, viscous friction, and surface tension related forces. The importance and order of magnitude of each force depends on the process parameters (such as laser beam power, travel speed, and spot area) and material [16]. Surface tension induces at least two forces at the interface between liquid and gas phases. The first component is called the capillary force which acts normal to the interface. It results from the cohesive forces between molecules that cause the surface of a liquid to contract to the smallest possible surface area. Molecules on the surface are pulled inward to the melt pool by cohesive forces, reducing the surface area. Molecules deeper inside the liquid experience zero net force, since they have similar neighbour molecules on all sides. During the melting process the free surface of the melt pool may be deformed due to the capillary force, but this effect may also be weakened by the viscous force. The capillary number can be used to evaluate the relative strength of the viscous and surface tension force as:

$$Ca = \frac{\mu U}{\sigma} \tag{2.1}$$

where σ is the value of the surface tension at the melting point, U is the characteristic velocity inside the melt pool, and μ is the viscosity. The typical value of the Ca number in conduction mode LBW is about 0.1 [17]. A Capillary number approaching zero indicates large surface tension effects. The second force induced by the surface tension force is called Marangoni force (or thermocapillary force). Its direction is tangential to the interface. The Marangoni force arises from variations of surface tension with respect to temperature. The significance of the Marangoni force with respect to viscous friction force can be expressed by the Marangoni number:

$$Ma = \frac{(\frac{d\sigma}{dT})L_c\Delta T}{\mu\alpha_T} \tag{2.2}$$

where $(\frac{d\sigma}{dT})$ is the variation of surface tension with temperature, L_c is the characteristic length, α_T is the thermal diffusivity, and ΔT is the characteristic temperature difference. The order of magnitude of the Marangoni number inside a melt pool is 10^2-10^5 [16]. It is also worth mentioning that the sign of the value $(\frac{d\sigma}{dT})$ determines the direction of the Marangoni force which is always from the lower surface tension to the higher one [16].

Another force in a melt pool is the buoyancy force which arises from the variations of liquid density with respect to the temperature. The magnitude of the buoyancy force is much less significant compared to the Marangoni force and can be quantified by the Bond number:

$$Bo = \frac{Ra}{Ma} = \frac{\rho g \beta L_c^2}{\left(\frac{d\sigma}{dT}\right)} \tag{2.3}$$

where Ra is the Rayleigh number, ρ is the alloy density at the melting temperature, g is the gravitational acceleration, and β is the thermal expansion coefficient of the liquid alloy. The order of magnitude of the Bo number inside a melt pool in conduction mode welding is about 10^{-4}-10^{-3} [16].
Heat and mass are transported by means of convection, conduction, or diffusion within the melt pool. The ratio of convective to conductive or diffusive transport is given by the Peclet number:

$$Pe = \frac{U L_c}{\alpha_T} \tag{2.4}$$

The order of magnitude of the Peclet number inside a melt pool is 10-10^2, which means that the convection effect is dominant compared to the conduction or diffusion [16].
Several attempts have been made in order to determine whether a melt pool is turbulent or laminar, but this question is still open. For free surface flow the Marangoni Reynolds number which could give a criterion about this transition is defined as [16]:

$$Re = \frac{\rho L_c \Delta T \left(\frac{d\sigma}{dT}\right)}{\mu^2} \tag{2.5}$$

The order of magnitude for the Marangoni Reynolds number inside a melt pool can reach as high as 10^5 corresponding to a highly unstable and a violent free surface flow. This has been observed experimentally by Zhao et al. [18]. The effect of turbulence is dependent on the process parameters, therefore it has not been agreed on a specific criterion by which one can distinguish between laminar and turbulent melt pool. Theoretically, considering the effect of turbulence on the melt pool thermo-hydrodynamic behaviour can be justified by two reasons. First it causes an increase in the effective thermal conductivity of the material which enhances the heat conduction. Second, it increases the effective viscosity of the material which suppresses the convection inside the melt pool. These two effects cause a decrease in the melt pool depth compared to when laminar flow is dominant (see Figure 2.4).
In order to account for the effect of turbulence in the melt pool, in some research works an artificial value called "enhancement factor" was used [16, 19,

20]. The purpose of defining the enhancement factor was to increase the effective thermal conductivity and/or viscosity which consequently led to the decrease in the melt pool size and made the numerical results match against the experimental data. In another approach turbulence modelling methods (mainly the k-epsilon model [21]) were used by some researchers [22–24]. Although, the obtained numerical results were satisfactorily matched against experimental data, these works lack proper justifications and logical explanations about why the flow was turbulent and what the turbulence criterion was. Kidess et al. [25, 26] studied a more sophisticated approach to investigate the effect of turbulence on the melt pool flow and temperature fields. They assumed that the flow field was turbulent when the Marangoni Reynolds number was higher than 600. They used both large eddy simulation (LES) [27] and direct numerical simulation (DNS) [28] to study the role of turbulence and its importance in the welding process.

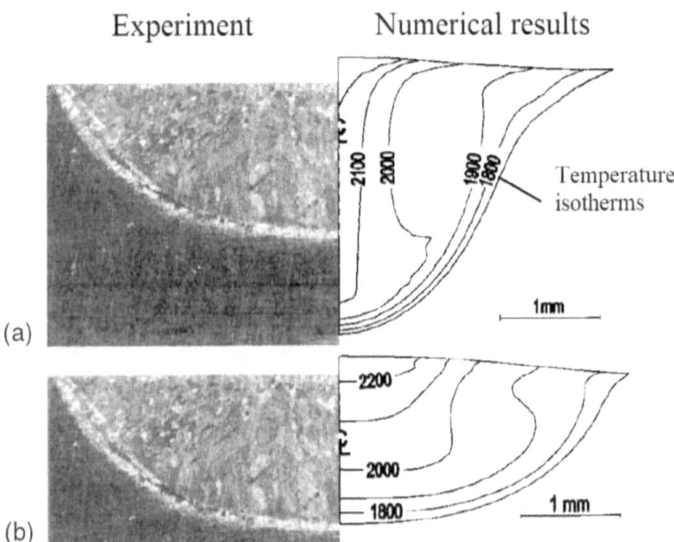

Figure 2.4: Categorization of single melt tracks (a) laminar, (b) turbulent [29].

Although their results showed the existence of transitional flow in the melt pool, they used some assumptions about the characteristic velocity and temperature difference inside the melt pool which were dependent on the process parameters in their cases and cannot be generalized to all other cases. Another point was that they did not mention any criterion at which the flow field could be considered as turbulent. In another approach, Chakraborty [30–32] used a scaling analysis in order to identify different flow regimes inside a melt pool and constructed a regime diagram for some specific materials (see Figure 2.5). He found that when the liquid metal Prandtl number (Pr) was very small (in the order of 0.01) the flow regime inside the melt pool was governed by diffusion and molten pool convection had small contribution to the weld pool morphology. On the other hand, at higher Prandtl number convection played a crucial role in determining the molten pool morphology. It was also found that the role of turbulence became important on the momentum transport inside the melt pool, however the thermal transport got marginally affected due to high thermal diffusivity of molten metals with very small Prandtl number.

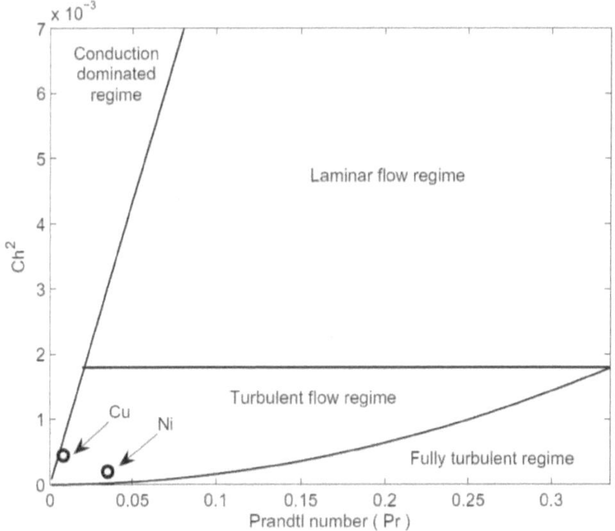

Figure 2.5: Molten pool convection in copper and nickel in the regime diagram for thermal transport [30].

The above discussion shows that there is no general agreement on the existence of turbulence in the melt pool. In this book it is assumed that the flow field is laminar. The obtained numerical results showed that such assumption was not far from reality for the test cases investigated in this study, since the deviations between the predicted results and experimental data for melt pool geometry were in acceptable range (below 10%).

Chapter 3

Introduction to the research questions

In this chapter two major factors corresponding to the research questions are discussed. First, the importance of the laser beam profile in welding is presented giving the relevant knowledge and research background. Then the effect of laser absorption is described. It will be shown that the effect of variable laser absorption has been neglected or too much simplified in the previous research works.

3.1   Laser beam profile

One of the substantial factors that significantly affect the melt pool thermo-hydrodynamic behaviour and geometry is the laser power density distribution (laser beam profile) on the surface of the working material. Among the various types of laser beam profiles, the Gaussian and top-hat distributions have the most applications. Nevertheless, during the last years it has been found that properly manipulating the laser beam profile might result in more stable welding and also improvement of the material micro-structure [33–36]. In numerical simulation, calculating the laser beam profile in a more realistic way would lead to more accurate results against the experimental data [37, 38].

Recently, Ayoola et al. [39] experimentally investigated the effect of different beam profiles on the bead geometry in the conduction mode welding. These authors changed the tilt angle of the laser beam, while keeping the power constant, in order to change the beam elongation along the welding direction (see Figure 3.1). Such beam manipulation at a constant laser power changes three important parameters. First, the beam spot area may change by changing the beam elongation, as a result the beam power density changes. Secondly, by changing the spot size along the welding direction the interaction time, which is the time of interaction between laser and working material along the welding direction, changes. Thirdly, the beam energy density which includes a combined effect of laser power density and interaction time may change by shaping the laser beam.

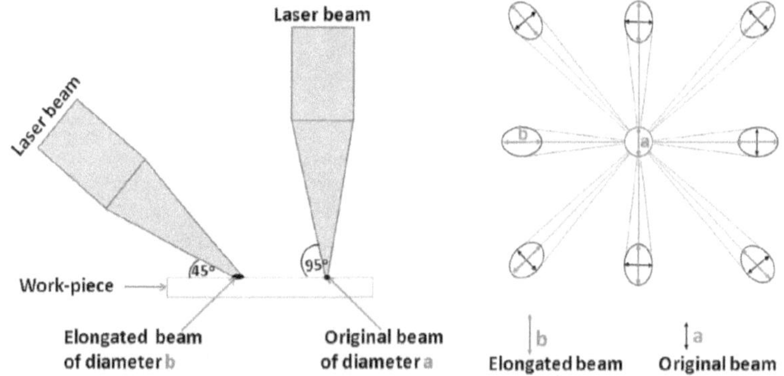

Figure 3.1: Schematic representation of beam inclination and its effect on the projected spot size [39].

The relevant parameters regarding the laser beam shaping are defined as [12]:

$$\text{Laser power } (P) = \iint_{A_s} P_d(x,y)\, dx\, dy \tag{3.1}$$

$$\text{Power density } (P_d) = \frac{P}{A_{\text{beam}}} \tag{3.2}$$

$$\text{Interaction time } (t_i) = \frac{L}{U_{\text{laser}}} \tag{3.3}$$

$$\text{Energy density } (E_d) = P_d\, t_i \tag{3.4}$$

$$\text{Specific point energy } (E_{sp}) = E_d\, A_{\text{beam}} \tag{3.5}$$

where A_{beam} is the laser spot area, L is the length of the beam spot along the welding direction, U_{laser} is the laser travel speed, x and y are the coordinates in the plane of the laser spot. Ayoola et al. [39] found that the beam spot spatial energy distribution had significant influence on the weld bead geometry. Their

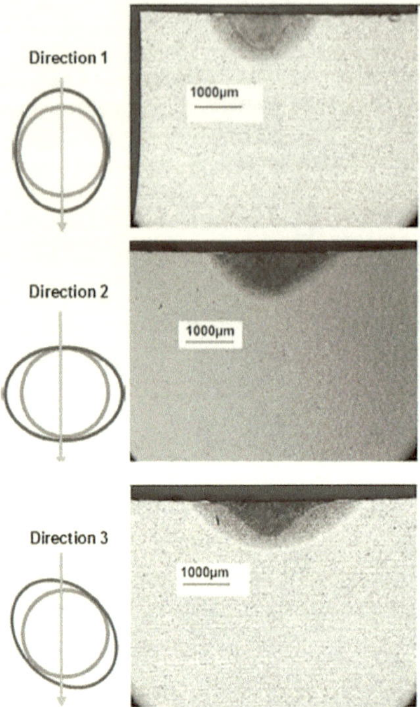

Figure 3.2: Bead-on-plate welds produced with different beam profiles [39].

results showed that the elongation of the laser beam in the welding direction resulted in deeper welds, whereas the elongation of the laser beam in the transverse direction resulted in wider welds. (see Figure 3.2).

Rasch et al. [33] studied the effect of different laser beam profiles on the melt pool stability and geometry in conduction mode welding. Examples of beam shapes that could be produced by their optics system are shown in Figure 3.3. They applied the different beam shapes to bead-on-plate welding and found that the effect of a laser beam with very high peak intensity did not necessarily lead to lower stability in welding. On the other hand, the laser beam profile itself played the dominant role in the stability of the process.

Relevance to the research question RQ1

Literature survey shows that the research on the effect of the laser beam shape on the melt pool geometry and thermo-hydrodynamic behaviour is mainly experimental. In addition, these experimental works do not give access to the ef-

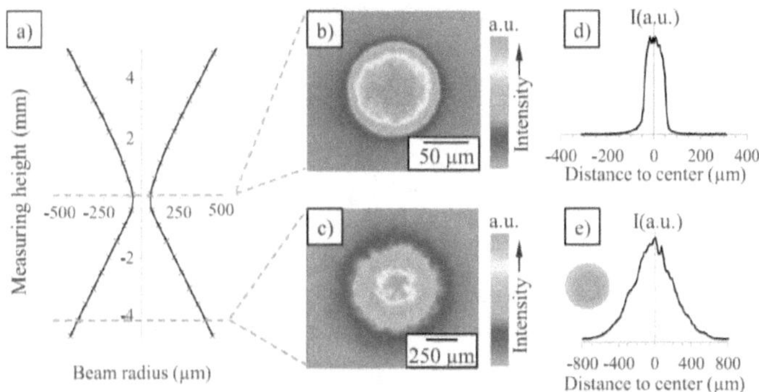

Figure 3.3: (a) Beam caustic (b) beam profile in the focal plane (c) defocussed beam profile (d) cross- section in the focal plane (e) cross-section through defocussed beam [33].

fect on the temperature and flow fields inside the melt pool for different beam shapes. Numerical simulation can provide this complementary information. Therefore, it can provide knowledge about the effect of the beam shape on the melt pool thermal and flow fields and fill a knowledge gap in this field. In this research work three types of laser beam profiles are used: top-hat, Gaussian, and elliptical:

- Top-hat power distribution:

$$\dot{q}_{\text{laser}}(r) = \begin{cases} \dfrac{\eta P}{\pi r_{\text{beam}}^2} & if \quad r \leq r_{\text{beam}} \\ 0 & if \quad r > r_{\text{beam}} \end{cases} \tag{3.6}$$

- Gaussian power distribution:

$$\dot{q}_{\text{laser}}(x,y) = \frac{2\eta P}{\pi r_{\text{beam}}^2} \exp\left[\frac{-2}{r_{\text{beam}}^2}(x^2 + y^2)\right] \tag{3.7}$$

- Elliptical power distribution:

$$\dot{q}_{\text{laser}}(x,y) = \frac{2\eta P}{\pi a b} \exp\left[-2(\frac{x}{a})^2 - 2(\frac{y}{b})^2\right] \tag{3.8}$$

where $\dot{q}_{\text{laser}}$ is the surface heat flux coming from the laser beam, η is the common notation for the so-called absorptivity coefficient used in conduction mode LBW, $r = \sqrt{x^2 + y^2}$ is the radius, and r_{beam} is the beam radius. It is worth-mentioning that, in the experiments, the beam radius is defined so that the amount of energy inside the corresponding area is 85% of the total laser power. The two parameters a and b are the ellipse radii along and transverse to the welding direction, respectively. The above-mentioned heat sources are moved at the travel speed U_{laser} along the welding direction.

Numerical simulations considering these different laser beam power density distributions are performed to address the first research question (RQ1): How do different power density distributions in the laser beam spot modify the process physics during LBW in conduction mode? Based on the state-of-the-art knowledge in this field, which is developed in the appended Paper 1 and thus not repeated here, RQ1 can be developed into the following sub-questions:

- **RQ1.1**: How does the flow pattern in a melt pool change (apart from being rotated) when a Gaussian-elliptic spot profile is oriented along different directions?

- **RQ1.2**: How different is the flow pattern when using a Gaussian profile rather than its elongation along some direction?

- **RQ1.3**: How does the thermocapillary flow change when lowering the power density with the different beam shapes?

These questions are addressed in the first appended journal article. It will be shown that the flow pattern inside the melt pool is considerably affected by using each beam profile mentioned above, that is why Ayoola et al. [39] observed different melt pool sizes by manipulating the beam profile at a constant laser power.

3.2 Laser beam absorption

In this section the information required to model the laser beam absorptance is provided. This information, gathered from various sources focusing on specific aspects of the problem, is combined in order to include the desired physics in the proposed model.

There are different notations for the laser absorptivity depending on authors and fields. The notation η is used in the field of welding and called absorptivity coefficient, while A is used in the field of optics and is called absorptance. This research work in the field of welding will make use of models developed within the field of optics. Therefore, in this study the fraction of beam energy absorbed by the material is called absorptivity coefficient when referring to the traditional LBW approach and absorptance when using the model proposed in this study based on optics.

Laser produces monochromatic, coherent, and collimated light that can be described by both electric and magnetic waves as illustrated in Figure 3.4. The electric ($\vec{E}$) and magnetic ($\vec{H}$) waves are always perpendicular when propagating through vacuum or through a medium. The electromagnetic wave can be characterized by its frequency, wavelength, and amplitude. Depending on the application different wavelengths may be used. In the current research work, a laser beam with the wavelength of 1.06 μm is used for welding of Ti-6Al-4V.

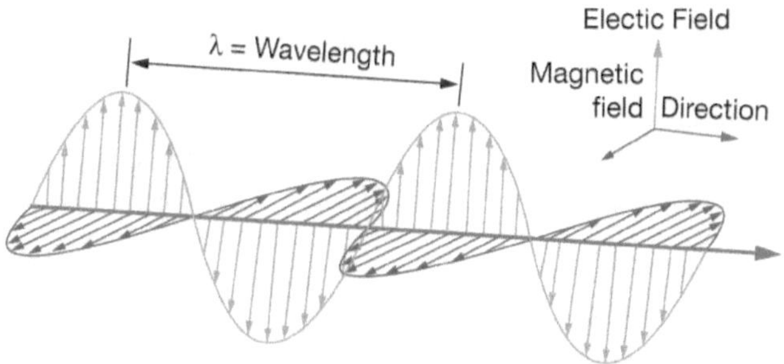

Figure 3.4: Schematic of an electromagnetic field [40].

To understand how to model the absorptivity, a study of electromagnetism and optics was conducted. Some elements that are not described in the appended papers are given here. Maxwell, for the first time, combined together the phenomena in electrostatics, magnetism, and electric current discovered by Gauss, Faraday, and Ampére which were considered to be separate and independent in the form of four differential equations for the electromagnetic field. These equations, for simplicity in general form for vacuum, can be presented as follows [41]:

$$\nabla . \vec{E} = \frac{\rho_c}{\epsilon_0} \tag{3.9}$$

$$\nabla . \vec{H} = 0 \tag{3.10}$$

$$\nabla \times \vec{E} + \frac{\partial \vec{H}}{\partial t} = 0 \tag{3.11}$$

$$\nabla \times \vec{H} = \mu_0 \vec{J}_\epsilon + \epsilon_0 \mu_0 \frac{\partial \vec{E}}{\partial t} \qquad (3.12)$$

where ρ_c is the charge density (in C/V), ϵ_0 is the permittivity of free space ($= 8.854 \times 10^{-12} Nm^2/C^2$), t is the time, μ_0 is the permeability of free space ($= 4\pi \times 10^{-7} Tm/A$), and $\vec{J}_\epsilon$ is the electric current density vector (in A/m^2). The free-space permittivity and the permeability are natural constants which are related to the velocity of light in vacuum by:

$$c = \frac{1}{\sqrt{\mu_0 \epsilon_0}} \qquad (3.13)$$

Each of the Maxwell's equations has a specific interpretation. It is useful to understand them in order to establish the boundary conditions at the interface between two media (gas and metal). Such boundary conditions are used in the appended papers where the absorptance model is obtained.
According to Eq.(3.9) the divergence of the electric field is a function of charge density (derived from Gauss' electric law). Eq.(3.10) states that the divergence of the magnetic field in a closed loop is zero (derived from Gauss' magnetic law). Eq.(3.11) is based on the Faraday's law of induction and states that a time-varying magnetic field induces an electric field. Finally, Eq.(3.12) states that a closed loop of magnetic field lines will exist in the presence of a current and/or time varying electric field. Another useful relation that can be used to simplify the Maxwell's equations is called the continuity equation. It expresses the conservation of charge and is presented as:

$$\nabla \cdot \vec{J}_\epsilon = -\frac{\partial \rho_c}{\partial t} \qquad (3.14)$$

This equation states that if the current vector diverges there must be a local decrease in the charge density with time as the current carries charge away. The above-mentioned Maxwell's equations are valid for vacuum which, of course, can be revised in a media and particularly for conductors such as metals. When expressed in a media the Maxwell's equations need to be supplemented with closure relations. The choice of closure relations is discussed in the appended Paper 3 and therefore not repeated here.
The charge density in a material is the summation of the free and the bound charge densities. The free charge density is related to the electric field as: $\vec{J}_\epsilon = \sigma_\epsilon \vec{E}$, where σ_ϵ is the electrical conductivity. Combining this definition with Eq.(3.9) and Eq.(3.14), yields:

$$\frac{\partial \rho_c}{\partial t} = -\sigma_\epsilon \nabla.\vec{E} = -\frac{\sigma_\epsilon}{\epsilon_m}\rho_c \tag{3.15}$$

where ϵ_m is the permittivity of the medium. The characteristic space scale in thermo-fluid problems is larger than the Debye screening length, implying that the positively charged ions are screened by the conduction electrons. Therefore at the thermo-fluid scale $\rho_c \approx 0$ and one can obtain (for conductors):

$$\nabla.\vec{E} = 0 \tag{3.16}$$

It is important to notice that when $\rho_c = 0$ then $\nabla.\vec{J_\epsilon} = 0$, but $\vec{J_\epsilon} \neq 0$, because a current can still exist in an electrically neutral conductor as the conduction electrons move relative to stationary ions. Another simplification is that the conductor is assumed to be homogeneous and isotropic, as a result, the variations of permittivity and permeability with respect to the direction are ignored. With closure relations and these simplifications the Maxwell's equations in conducting media are reduced to:

$$\nabla.\vec{E} = 0 \tag{3.17}$$

$$\nabla.\vec{H} = 0 \tag{3.18}$$

$$\nabla \times \vec{E} + \frac{\partial \vec{H}}{\partial t} = 0 \tag{3.19}$$

$$\nabla \times \vec{H} = \mu_m \sigma_\epsilon \vec{E} + \epsilon_m \mu_m \frac{\partial \vec{E}}{\partial t} \tag{3.20}$$

where μ_m is the permeability of the medium. One can take the curl of Eq.(3.19) and use the relationship $\nabla.\vec{E} = 0$ to separate electric and magnetic fields and obtain the following equation:

$$\nabla^2 \vec{E} = \mu_m(\sigma_\epsilon \frac{\partial \vec{E}}{\partial t} + \epsilon_m \frac{\partial^2 \vec{E}}{\partial t^2}) \tag{3.21}$$

If a similar procedure is applied to Eq.(3.20), the following equation can be obtained for the magnetic field:

$$\nabla^2 \vec{H} = \mu_m (\sigma_\epsilon \frac{\partial \vec{H}}{\partial t} + \epsilon_m \frac{\partial^2 \vec{H}}{\partial t^2}) \qquad (3.22)$$

In order to solve Eq.(3.21), which is a wave equation with addition of an extra first derivative term, some more assumptions have to be made. The solution can be a monochromatic, plane, and coherent wave (like laser beam). A monochromatic wave is a wave with single wavelength and frequency. Moreover, the plane wave is a constant-frequency wave whose wave-fronts are infinite parallel planes of constant amplitude normal to the phase velocity vector. Additionally, two waves are coherent if their frequency and waveform are identical and their phase difference is constant. The fields for such wave can be presented as (assumed to propagate in the z direction):

$$\vec{E}(z,t) = \vec{E}_0 e^{i(\vec{k}_\lambda z - \omega_\lambda t)} \qquad (3.23)$$

$$\vec{H}(z,t) = \vec{H}_0 e^{i(\vec{k}_\lambda z - \omega_\lambda t)} \qquad (3.24)$$

where $\vec{k}_\lambda$ is the wave vector and ω_λ is the wave frequency. In a damping media like a metal, the wave vector $\vec{k}_\lambda$ is a complex parameter, which means that the resulting wave has both an oscillatory and an exponentially decaying factor. The wave vector $\vec{k}_\lambda$ is related to the complex refractive index of the material, $n_\epsilon = n'_\epsilon - i n''_\epsilon$, as:

$$|\vec{k}_\lambda| = k_\lambda = \frac{\omega_\lambda}{c} n_\epsilon \qquad (3.25)$$

where the wave frequency and wavelength are related as $\omega_\lambda = \frac{2\pi}{\lambda}$. In the complex refractive index, the real part (n'_ϵ) is called the refractive index, while the imaginary part (n''_ϵ) is the extinction coefficient.

In laser welding a laser beam (modelled by Eqs. (3.23 and 3.24)) comes from a gas phase and impinges on the surface of a metal (in this research work Ti-6Al-4V). The laser beam may partially be transmitted, absorbed, and reflected. In metal workpieces as used in welding all the transmitted beam is absorbed by the material (no transparency). Moreover, it is considered that when a laser beam hits a surface only unidirectional reflection happens. In this way an incident plane wave remains planar after reflection. The transmittance of a beam is then governed by Snell's law (which describes the relationship between the incident and refraction angles), while the reflection of a beam is governed by law of reflection (which states that in case of plane wave the incident and reflected angles are equal). These two laws in an arbitrary plane can be presented as:

- Law of reflection:

$$\theta_i = \theta_r \tag{3.26}$$

- Snell's law:

$$\frac{sin\theta_i}{sin\theta_t} = \frac{n_{\epsilon,2}}{n_{\epsilon,1}} = \frac{n'_{\epsilon,2} - in''_{\epsilon,2}}{n'_{\epsilon,1} - in''_{\epsilon,1}} \tag{3.27}$$

where θ_i, θ_r, and θ_t, are the incident, reflected, and transmitted angles respectively. The indices 1 and 2 refer to the different media on both sides of the interface. It is also worth mentioning that the Snell's law cannot hold when $\mid n_{\epsilon,1} \mid > \mid n_{\epsilon,2} \mid$. In such case the incident wave is totally reflected then there is no transmitted wave.

In order to investigate what portion of an incident wave (in another word what fraction of energy) is transmitted (=absorbed) and reflected when hitting a metal surface, Fresnel's model is used along with Eqs. (3.26) and (3.27). This finally leads to the concept of laser beam absorption. A domain at which a plane electromagnetic wave hits an interface of two media with different permeabilities and permittivities is shown in Figure 3.5. Since the electromagnetic wave is transverse, the field incident onto the interface can be decomposed into two polarization components, one p-polarized, i.e., with the electric field vector inside the plane of incidence, and the other one s-polarized, i.e., with the electric field orthogonal to that plane.

The boundary condition at the interface for the electric and magnetic fields in case of p-polarization can be presented as:

$$E_i cos\theta_i + E_r cos\theta_r = E_t cos\theta_t \tag{3.28}$$

$$H_i - H_r = H_t \tag{3.29}$$

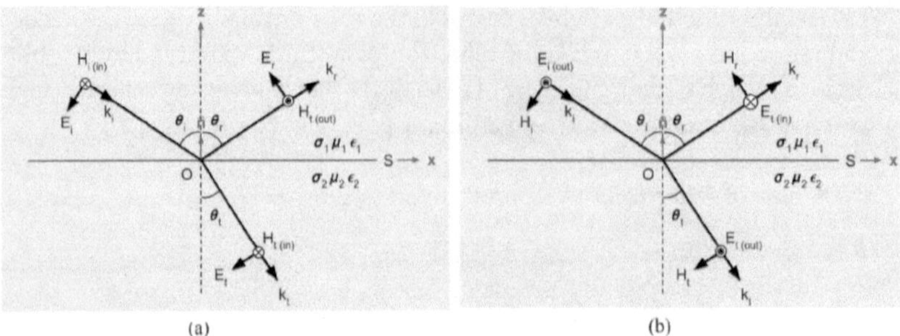

Figure 3.5: Incident, reflected and transmitted waves with: (a) p-polarization, (b) s-polarization [42].

With the help of the auxiliary relationship between electric and magnetic fields ($|\vec{H}| = \sqrt{\mu\epsilon}\,|\vec{E}|$, from Maxwell's equations) and Eqs. (3.13), (3.26), and (3.27) the above boundary conditions can be re-arranged as follows:

$$r_p = \left(\frac{E_r}{E_i}\right)_p = \frac{(n_{\epsilon,1}/\mu_1)\cos\theta_t - (n_{\epsilon,2}/\mu_2)\cos\theta_i}{(n_{\epsilon,1}/\mu_1)\cos\theta_t + (n_{\epsilon,2}/\mu_2)\cos\theta_i} \tag{3.30}$$

$$t_p = \left(\frac{E_t}{E_i}\right)_p = \frac{2(n_{\epsilon,1}/\mu_1)\cos\theta_i}{(n_{\epsilon,1}/\mu_1)\cos\theta_t + (n_{\epsilon,2}/\mu_2)\cos\theta_i} \tag{3.31}$$

The boundary conditions at the interface in case of s-polarization for the electric and magnetic fields can be presented as:

$$E_i - E_r = E_t \tag{3.32}$$

$$H_i \cos\theta_i - H_r \cos\theta_r = H_t \cos\theta_t \tag{3.33}$$

Similar to the case of p-polarization the above equations can be re-written for the s-polarization as follows:

$$r_s = \left(\frac{E_r}{E_i}\right)_s = \frac{(n_{\epsilon,1}/\mu_1)\cos\theta_i - (n_{\epsilon,2}/\mu_2)\cos\theta_t}{(n_{\epsilon,1}/\mu_1)\cos\theta_i + (n_{\epsilon,2}/\mu_2)\cos\theta_t} \tag{3.34}$$

$$t_s = (\frac{E_t}{E_i})_s = \frac{2(n_{\epsilon,1}/\mu_1)\cos\theta_i}{(n_{\epsilon,1}/\mu_1)\cos\theta_i + (n_{\epsilon,2}/\mu_2)\cos\theta_t} \tag{3.35}$$

Eqs. (3.30), (3.31), (3.34), and (3.35) are called Fresnel equations. This set of equations gives information about the portion of reflected and/or transmitted plane electromagnetic wave after hitting a surface under the discussed conditions and assumptions. The amount of energy that is being carried by a monochromatic plane electromagnetic wave is calculated as:

$$I_w = \frac{1}{\mu}(\vec{E} \times \vec{H}) \tag{3.36}$$

The amount of energy that is transmitted into the material (in other word is absorbed under the described conditions) can be calculated as:

$$A_{\text{p-polarized}} = (\frac{E_t}{E_i})_p^2 = t_p t_p^* \tag{3.37}$$

$$A_{\text{s-polarized}} = (\frac{E_t}{E_i})_s^2 = t_s t_s^* \tag{3.38}$$

where t_p^* and t_s^* are the complex conjugates of t_p and t_s respectively. Then for a circularly polarized wave as in laser beams used in material processing the average of surface transmittance (absorption) are calculated as:

$$\eta = A = \frac{A_{\text{p-polarized}} + A_{\text{s-polarized}}}{2} \tag{3.39}$$

The above-mentioned formulation has been used in many research works particularly in keyhole mode welding applying further mathematical and physical simplifications when accounting for the effect of the beam angle and surface curvature [43–46]. In available published research works the refractive indices were assumed to be constant and set depending on the material and application. It implies that the absorptivity varied only based on the incident beam angle. There are also some experimental works in which the laser absorptivity is either measured or formulated with respect to the surface temperature and the laser wavelength. Literature survey shows that there are two most-common relations proposed by Hagen-Ruben and Bramson for predicting the laser absorptivity with respect to the wavelength and temperature in metals. Xie et al. [47] proposed a function based on Hagen-Ruben relationship for different materials and lasers as:

$$\eta(\lambda, T) = 0.365(\frac{1}{\sigma_\epsilon(T)\lambda})^{0.5} \tag{3.40}$$

where λ is the incident beam wavelength (in m), T is the surface temperature (in K), and σ_ϵ is the material electrical conductivity (in $(\Omega m)^{-1}$). Webber et al. [48] used a more accurate relation based on Bramson model:

$$\eta(\lambda, T) = 0.365(\frac{1}{\sigma_\epsilon(T)\lambda})^{0.5} - 0.0667(\frac{1}{\sigma_\epsilon(T)\lambda}) + 0.006(\frac{1}{\sigma_\epsilon(T)\lambda})^{1.5} \tag{3.41}$$

These two relationships were derived based on Drude model with different levels of simplifications [49]. Although these equations give good prediction of the absorptivity, they are derived for a beam perpendicular to the surface material (normal incident). Therefore, they cannot be used for other incident angles. The variations of the laser absorptivity with respect to temperature and laser beam wavelength for Ti-6Al-4V and normal beam incidence based on Hagen-Ruben and Bramson equations are shown in Figure 3.6. A large difference, particularly at the low wavelength range, can be seen in the results. This is a critical issue since in the present research work a laser beam with a wavelength of 1.06 μm is used.

There are also some research works in which the laser absorptivity was measured experimentally (mainly for temperatures below the melting point). The most critical problem with these experiments is that the measured values are only for a normal beam incidence. Moreover, the experimental works were limited to particular process conditions. Yang et al. [50] experimentally obtained a value of 0.34 for the laser absorption coefficient during welding of Ti-6Al-4V over a wide range of temperature ($773 < T < 1673$ K). They used a Nd:YAG laser system in their research work. Hagqvist et al. [51] measured the radiative emissivity ($\epsilon(\lambda, T) = \eta(\lambda, T)$), based on Kirchhoff's law) of Ti-6Al-4V in the temperature range between 750 K to 1550 K for the wavelength of $\lambda = 1.06\mu m$. They reported that the radiative emissivity values varied from 0.32 to 0.36 depending on the temperature. However, during welding the laser beam mainly interacts with the liquid metal surface. Therefore the temperature range is larger than the ones in these measurements. In another experimental work, Gonzalez-Fernandez et al. [52] measured the radiative emissivity over the temperature range of 550 to 1150 K and wavelength of $1.5\mu m$ to $22\mu m$. The reported values of absorptivity at $\lambda = 1.5\mu m$ varied approximately between 0.28 to 0.35 when the temperature changed from 1000 K to 500 K. Table

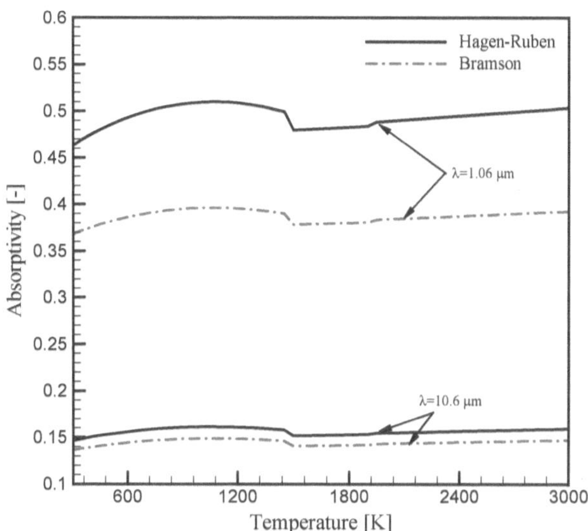

Figure 3.6: Variations of Ti-6Al-4V absorptivity with respect to laser wavelength and surface temperature based on two different relations from Hagen-Ruben [47] and Bramson [48].

3.1 presents a comparison between different values of absorptivity reported by previous research works through experimental and semi-empirical techniques for a normal beam incidence and Ti-6Al-4V as the working material.

Table 3.1: Comparison between different references for determination of the laser absorptivity coefficient for a normal beam and Ti-6Al-4V as the working material at room temperature.

$\lambda(\mu m)$	[47]	[48]	[53]	[50]	[54]	[55]	[56]
0.80	0.533	0.409	-	-	0.52	-	-
1.06 (Nd:YAG laser)	0.463	0.368	-	0.34	-	-	0.50
3.8	0.244	0.214	0.235	-	-	-	
10.6 (CO_2 laser)	0.146	0.136	0.135	-	-	0.11	-

It is obvious that the deviations between almost all the results are not small. Furthermore, these data were obtained in temperature ranges that are below the temperatures of interest in metal fusion.

Relevance to the research question RQ2 and RQ3
In conduction mode LBW, the effect of the surface curvature was always neglected in previous research works. Moreover the effect of temperature on the refractive index and thus on the absorptivity was not considered. In this research work, these two effects are used in order to develop an absorptance model applied in conduction mode welding to address the following research questions:

RQ2: How can the absorptivity of laser beam energy by a metal alloy be predicted when simulating LBW in conduction mode with CFD?

- **RQ2.1**: How relevant is it to model the absorptivity of laser beam energy with regards to melt pool free surface geometry in LBW in conduction mode?

- **RQ2.2**: What are the assumptions and limitations for modelling the absorptivity of laser beam energy in LBW?

- **RQ2.3**: How important are the model variables for the process?

RQ3: How does a predicted absorptance affect the process physics during LBW compared to a constant absorptivity coefficient?

- **RQ3.1**: How is it possible to determine an average value for laser absorption which leads to the identical results as in the local prediction approach in conduction mode welding of Ti-6Al-4V?

- **RQ3.2**: How does the non-uniformity of absorption affect the flow field inside melt pool?

Chapter 4

Numerical method

In this chapter the developed numerical solver for the simulation of LBW, fundamental assumptions, calculation procedure for material properties, and solver settings for obtaining stable solutions are discussed.

4.1 Free surface modelling

The formation of the melt pool during welding can be treated as a free surface problem. It consists of three different phases: solid and liquid phases of the workpiece, and the top gas phase. Regardless of the method employed, there are three essential features needed to properly model the free surface of gas and liquid melt pool: (1) a method is needed to describe the shape and location of the free surface, (2) an algorithm is required to evolve the shape and location of free surface with time, (3) free-surface boundary conditions must be applied at the interface [57, 58]. There are two main categories of interface solvers. The first category involves grid motion and deformation, while the second group is based on fixed mesh. In this regard, there are three main methods to deal with the free surface problem based on the above-mentioned groups: (i) interface fitting method, (ii) interface capturing method, and (iii) interface tracking method [59]. Figure 4.1 shows the schematic of each method.

In the surface fitting methods, the interface is tracked by attaching it to a surface mesh which is forced to move with the interface. This method can be considered as a Lagrangian mesh method. The surface capturing methods, like Marker and Cell (MAC) [60], Volume of Fluid (VOF) [61], Level Set (LS) [62], and Diffuse Interface (DI) (or phase field) [63], use an indicator function to mark the fluids on both sides of the interface. Therefore the mesh remains fixed and the method is considered as an Eulerian mesh method. In the surface tracking methods, like Glimm's front tracking method [64], the interface is represented and tracked explicitly by special marker points. In this method the grid also

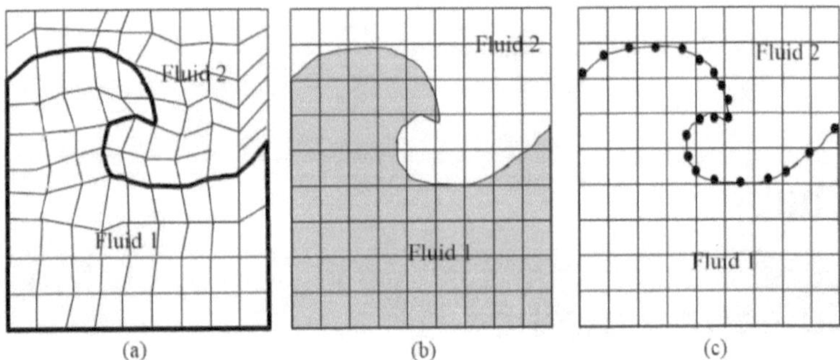

Figure 4.1: Free surface modelling schemes; (a) interface fitting method, (b) interface capturing method, (c) interface tracking mehtod [59].

remains fixed, therefore it can be considered as a hybrid Eulerian-Lagrangian mesh method [59]. Interface fitting methods can be very efficiently applied in the study of shaping processes of viscous materials. However, these methods become inaccurate in the case of less viscous materials that undergo large deformations. Mesh regenerations are necessary when mesh cells and elements become highly skewed, to prevent computational failures. This can make the interface fitting scheme and other moving mesh techniques become very complicated and inefficient [59]. Interface tracking and interface capturing methods avoid the grid-related problems associated with interface fitting methods by employing a fixed grid and defining the location of the liquid and gas regions relative to it. In general, interface capturing methods can handle complex free surfaces more easily, but interface tracking methods give a more accurate description of the free surface because of their Lagrangian nature [59, 65].

Among the various interface capturing schemes, MAC, VOF, and LS attracted more attentions. In MAC, hundreds of massless marker particles are added to the fluid. These particles are then advected in Lagrangian sense using the average of Eulerian velocities in their vicinity. In the VOF method, the volume fractions of fluids are used to mark the fluid regions. The volume fractions are convected through the flow domain by solving a scalar convection equation. In the level set method, a smooth distance function is used as the indicator function for the interface [59, 65]. Although the VOF method is very efficient, simple to implement, and able to handle complex interfaces, it leads to the smearing of the interface, whereas LS can predict the interface sharpness more accurately. Therefore, for flow involving surface tension, VOF may inaccurately predict the surface curvature and surface tension. However, VOF is superior to LS in that it guarantees a much better mass conservation [16]. To overcome the smearing caused by the volume fraction convection special consideration needs

to be taken to minimize the numerical diffusion. The following approaches are used to accomplish this: (1) free surface reconstruction using line techniques [66,67]; (2) high-resolution differencing schemes [68,69].
In this research work the VOF method is used for predicting the location and evolution of the melt pool free surface, as its performance has already been proved for this kind of problem [16]. The detailed information about this method is explained in the following section.

4.2 Governing equations and assumptions

In this research work the following assumptions are considered when determining the governing equations:

- The flow field is transient, three dimensional, and laminar [70,71].

- The fluid is mechanically incompressible. For the metal in the liquid state the variation of the density with temperature is small compared to the metal density at the melting point. These thermal variations of the fused metal density are thus included through the Boussinesq approximation [72].

- The diffusion of alloying elements is neglected and the material properties vary only with temperature.

- Since this work concerns only conduction mode welding with low laser power density, the metal vaporization is negligible.

- The fluids are immiscible and the velocity difference between fluid phases is negligible. This assumption is valid for the case of free surface problem with non inter-penetrating media [73,74].

- Viscous dissipation is negligible, because the Prandtl number as well as the velocity magnitudes within the melt pool are low.

Based on the above-mentioned assumptions the governing equations (consisting of the continuity, volume fraction, momentum, and energy equations) and solution procedures are defined. The details about these governing equations are provided in the appended papers.

4.3 Transport properties

In this project the working material is Ti-6Al-4V, which is also sometimes called Ti64. The chemical composition of Ti-6Al-4V, in %wt, is presented in Table

4.1. It is a Titanium alloy with a high strength-to-weight ratio and excellent corrosion resistance. It is one of the most commonly used Titanium alloys and is applied in a wide range of applications where low density and excellent corrosion resistance are necessary such as aerospace industry, and bio-mechanical applications [75]. The main constituents of this alloy are paramagnetic with weak magnetism. This alloy has therefore negligible magnetism.

Table 4.1: Chemical composition of Ti-6Al-4V in %wt [76].

Al	Fe	Ti	V	H	O_2+N_2
5.5-6.7	0.03	90	3.5-4.5	0.0125	0.25

For simulation of LBW, the variations of some thermodynamic and transport properties of Ti-6Al-4V including density (ρ), thermal conductivity (k), viscosity (μ), specific heat capacity (c_p), latent heat of fusion (h_{sf}), radiative emissivity (ϵ), electrical resistance ($\rho_{electrical}$), and surface tension (σ) with respect to temperature (T) are needed. In the followings, the variations of the selected transport properties with respect to temperature are presented. The presented data are selected out of various experimental databases. Among these data the most approved ones are selected for each property.

Melting point: Since the working material, Ti-6Al-4V, is an alloy it does not have equal liquidus (T_l) and solidus (T_s) temperatures. The melting point of the alloy (T_m) is calculated as an average of the aforementioned temperature values [76]:

$$T_s[\mathrm{K}] = 1877, \quad T_l[\mathrm{K}] = 1923 \tag{4.1}$$

$$T_m[\mathrm{K}] = \frac{T_s + T_l}{2} = 1900 \tag{4.2}$$

Density: The available experimental data for the variations of density with respect to temperature for Ti-6Al-4V is shown in Figure 4.2. Although the density of solid Ti-6Al-4V varies with temperature, these variations are below 5% from room temperature to the melting point. Therefore these variations are neglected. On the other hand, the variation in the density of liquid Ti-6Al-4V is considered as a function of temperature through the Boussinesq approximation. This variation is important since it governs the buoyancy and gravity forces inside the melt pool. In this work the experimental data compiled by Mills [76] is selected (as used by many other researchers). For convenience these data were curve-fitted and then implemented into the solver. Based on this data the thermal expansion coefficient of the metal in liquid state, β, is assumed to be constant ($=6 \times 10^{-5}$ 1/K) [77]. This coefficient is used to express the density as a linear function of temperature in the Boussinesq method.

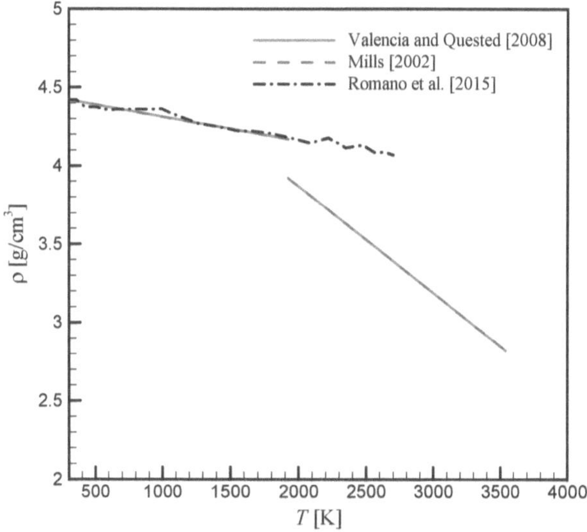

Figure 4.2: Variations of density of Ti-6Al-4V with respect to temperature [76, 78, 79].

Specific heat capacity: The specific heat capacity of Ti-6Al-4V at solid and liquid states is assumed to be a piece-wise function of temperature only. The available experimental data for the specific heat capacity of Ti-6Al-4V at different temperatures are shown in Figure 4.3. Since the experimental data compiled by Mills [76] is the most complete one and has been used in may other research works, it is used in this project as well.

Thermal conductivity: There are two well-known experimental databases for the variation of the thermal conductivity with respect to temperature. They can be seen in Figures 4.4 and 4.5. Both databases show two abrupt changes in the plots, one at the α-β phase transition and the other one at the solid-liquid

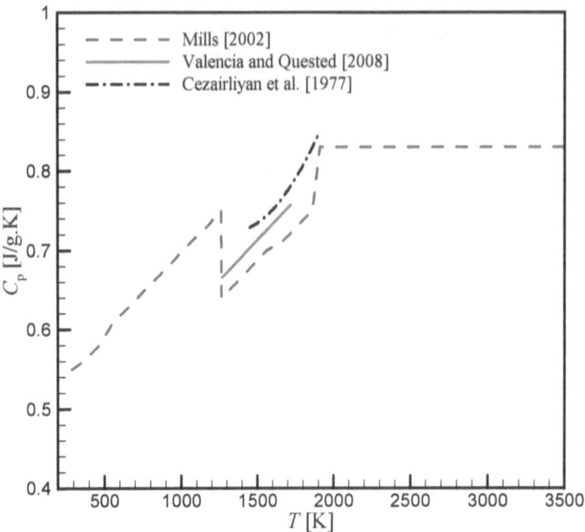

Figure 4.3: Variation of specific heat capacity of Ti-6Al-4V with respect to temperature [71, 76, 80].

phase change. Another point is that the variation of the thermal conductivity with respect to temperature has a piecewise linear trend. In this work the experimental data proposed by Boivineau et al. [81] are selected and the corresponding curve-fitting functions were added to the solver.

Surface tension: There is a well-approved experimental database for surface tension of Ti-6Al-4V as shown in Figure 4.6. In this work, based on this measurement, the surface tension that is defined in the liquid state is divided into two terms; the first one is a fixed value (σ_0) which is the surface tension of pure Titanium at the melting point, and the second one is its variations with respect to temperature ($\frac{d\sigma}{dT}$) [82, 83]:

$$\sigma[\text{N/m}] = \sigma_0 + \frac{d\sigma}{dT}(T - T_m) \tag{4.3}$$

$$\sigma_0[\text{N/m}] = 1.6, \quad \frac{d\sigma}{dT}[\text{N/m.K}] = -2.6 \times 10^{-4} \tag{4.4}$$

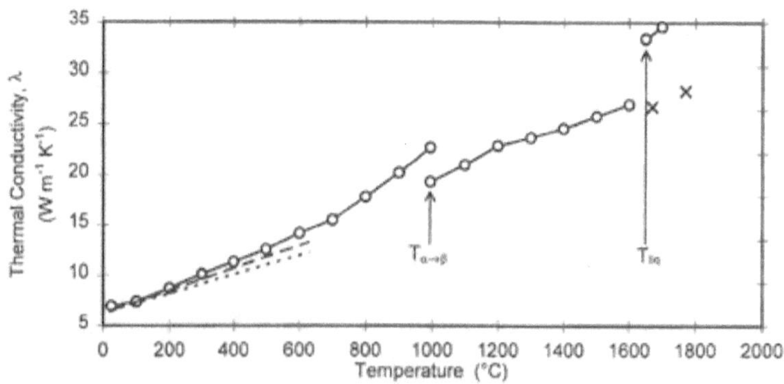

Figure 4.4: Variation of thermal conductivity of Ti-6Al-4V versus temperature [81].

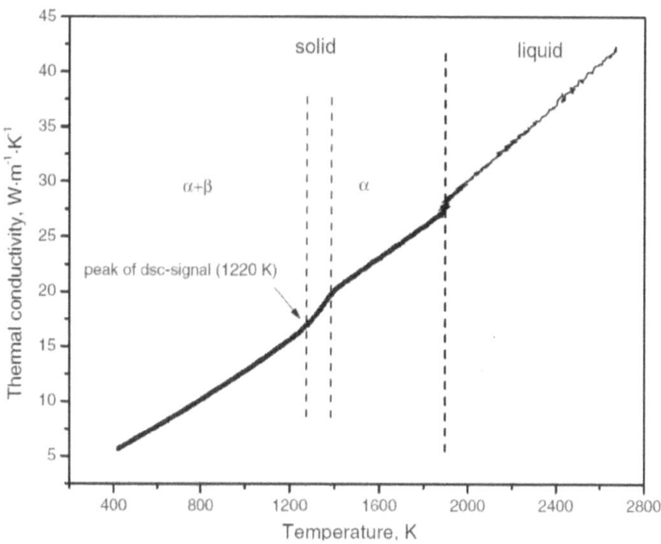

Figure 4.5: Variation of thermal conductivity of Ti-6Al-4V with respect to temperature [76].

The value of $(\frac{d\sigma}{dT})$ is considered to be constant since it is assumed that there is no surfactant in the solution.

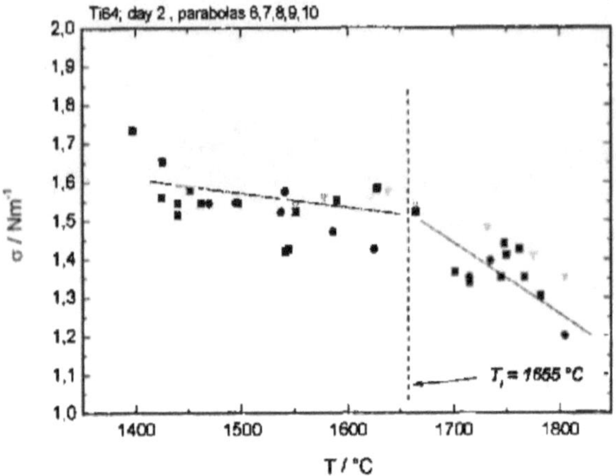

Figure 4.6: Variation of surface tension of Ti-6Al-4V with respect to temperature [84].

Liquid viscosity: The value of liquid metal viscosity is very rare in the literature, since its measurement seems to be very problematic. In this work it is assumed that the liquid metal viscosity is a constant, set based on the few available data [82, 83]:

$$\mu[\text{Pa.s}] = 0.0035 \qquad (4.5)$$

Latent heat of fusion: The value of h_{sf}, which is needed to calculate the heat released due to phase change, can be found as [76]:

$$h_{sf}[\text{J/kg}] = 2.86 \times 10^5 \qquad (4.6)$$

Radiative emissivity: Unfortunately, there is no generally-approved experimental database for the value of emissivity changes with respect to temperature. Nevertheless, a more common equation which has been used in some research works in the field of LBW for the total hemispherical emissivity is defined as [85]:

$$\epsilon = 0.1536 + 1.8377 \times 10^{-4}(T - 298) \ (300 \le T \le 3200, \text{K}) \tag{4.7}$$

Electrical resistivity: This parameter is used for the calculation of the laser beam absorption as discussed in Chapter 3. The experimental variations in electrical resistivity of liquid Ti-6Al-4V with respect to temperature can be fitted with the following equations as [81, 86, 87]:

$$\rho_{\text{electrical}}[\mu\Omega.\text{m}] = \begin{cases} 10^{-6}h^2 + 0.0011h + 1.7663 & (h < 850, \text{kJ/kg}) \\ 0.0001h + 1.7428 & (850 \le h, \text{kJ/kg}) \end{cases}$$

$$\tag{4.8}$$

4.4 OpenFOAM software

In the field of Computational Fluid Dynamics several commercial and free-source solvers were introduced during the last decades such as: STAR CCM+, ANSYS FLUENT, FLOW 3D, COMSOL, OpenFOAM, and CFX. Each of these solvers has its own advantages and drawbacks. In this project the Open-FOAM software, which is based on the Finite Volume Method (FVM), is used due to the following reasons:

- Complete access to the source code.

- No license costs.

- Friendly syntax for partial differential equations.

- Automatic parallelization of applications using OpenFOAM high-level syntax.

- Code customization according to problems.

- Commercial support and training provided by the developers.

In this work *interFoam* [88] which is available in the OpenFOAM software was used as the base solver and further developed. This solver solves the Navier-Stokes equations for two incompressible, Newtonian, and isothermal fluids using the VOF method in transient framework. The following new features were implemented into the original solver:

- Energy equation to include the non-isothermal cases.

- Phase change phenomenon.

- Temperature-dependent material properties.

- Marangoni force with the sharp surface force (SSF) model [89].

- Various laser beam power density distributions in relative motion with respect to the workpiece.

- Laser absorptance model as a function of temperature, surface curvature, and laser beam wavelength.

The details about the *interFoam* solver, solution procedure and methods are already provided by Svenungsson [90], therefore they are not repeated here. However, a short explanation about some relevant aspects of the solver are discussed below.

4.4.1 Finite Volume Method (FVM)

The OpenFOAM software uses the FVM cell-centred discretization of the domain and handles unstructured mesh data format based on the so-called face-addressing storage. The FVM was introduced into the field of computational fluid dynamics in the beginning of the seventies. From the physical point of view the FVM is based on balancing fluxes through control volumes. From the numerical point of view the FVM is a generalization of the Finite Difference Method (FDM) in a geometric and topological sense. The FDM is based on nodal relations for differential equations, whereas the FVM is a discretization of the governing equations in integral form. The FVM has two major advantages. First, it enforces conservation of quantities at discretized level, i.e. mass, momentum, energy remain conserved also at a local scale. Fluxes between adjacent control volumes are directly balanced. Second, finite volume schemes take full advantage of arbitrary meshes to approximate complex geometries [91]. The FVM method discretizes the problem in two ways:

- Spatial discretization

- Temporal discretization

Spatial discretisation requires the subdivision of the domain into a number of cells, or control volumes. The cells are contiguous (which means that they do not overlap with one another) and completely fill the computational domain. It follows that the boundary of the space domain is decomposed into faces connected to the cells next to them. The temporal discretization is done through integration over time on the general discretized equation. First, values at a given control volume at time t are assumed and then values at time $t + \Delta t$ are found. This method states that the time integral of a given variable is equal to a weighted average between current and future values.

4.4.2 Pressure-velocity coupling

There are two unknowns in the momentum equation, that are the velocity and pressure, therefore another equation is necessary to close the problem. Generally, the pressure is calculated with the help of the continuity equation together with the momentum equation. There are two widely used methods for dealing with pressure-velocity coupling, Semi Implicit Method for Pressure-Linked Equations (SIMPLE) [92] and Pressure Implicit with Splitting of Operators (PISO) [93]. In this research work the PISO algorithm is used as its superior performance compared to SIMPLE method in dealing with unsteady incompressible flows has been proved by many researchers [94, 95]. PISO algorithm generally gives more stable results and takes less CPU time. The solution procedure in PISO algorithm for an unsteady incompressible flow problem can be presented as [16]:

1. Set the boundary conditions.

2. Solve the discretized momentum equation to compute an intermediate velocity field.

3. Compute the mass fluxes at the cells faces.

4. Solve the pressure equation.

5. Correct the mass fluxes at the cell faces.

6. Correct the velocities on the basis of the new pressure field.

7. Update the boundary conditions.

8. Repeat from 3 for the prescribed number of times.

9. Advance the time and repeat from step 1.

4.4.3 Solution method for the volume fraction equation

In general solving the volume fraction equation may present two main difficulties, first numerical diffusion of the interface, and second unboundedness of the volume fraction value. In OpenFOAM the first difficulty is handled by introducing the numerical compression velocity. The second problem is solved by applying Multidimensional Universal Limiter for Explicit Solutions (MULES) method [88]. MULES is an iterative implementation of the Flux Corrected Transport (FCT) technique, used to guarantee boundedness in the solution of hyperbolic (here volume fraction equation) problems. This method computes a corrected flux between a high- and a low-order scheme solution with a weighting factor. The fundamental rule to determine the weighting factor is that the value of the net flux in a cell must be neither greater than a local maximum nor lower than a local minimum correcting the flux in order to not create new maximum or minimum [96].

4.4.4 Solver setting parameters

Several settings and adjustments must be done in order to obtain stable results with a desired numerical accuracy. These settings should be performed through some adjustable numerical parameters, schemes, and functions, which are accessible in three folders in each OpenFOAM simulation case namely *fvSoltion*, *fvSchemes*, and *controlDict*. These folders and the corresponding numerical parameters will be discussed in detail in the following sections [88].

4.4.5 Discretization schemes in *fvSchemes*

OpenFOAM software has various numerical schemes for terms, such as derivatives in equations, that are calculated during a simulation. The set of terms, for which numerical schemes must be specified, are subdivided in the *"fvSchemes"* dictionary into the categories below [88]:

- timeScheme: $\frac{\partial}{\partial t}$

- gradSchemes: gradient ∇

- divSchemes: divergence $\nabla \cdot$

- snGradSchemes: component of gradient normal to a cell face

- laplacianSchemes: Laplacian ∇^2

- interpolationSchemes: cell to face interpolations of values

As mentioned above, there are various schemes implemented in OpenFOAM to compute these operators. Here only the utilized or examined schemes are described [97–99].

- **Time derivatives**: For discretization of the first-order derivative terms with respect to time ($\frac{\partial}{\partial t}$), the *Euler* scheme which is an implicit first order method is used. This method showed more stability in the primary testing simulations compared to the *Crank Nicolson* scheme which is a second order method.

- **Gradient schemes**: For this term the *Gauss* method is used which specifies the standard finite volume discretization of Gaussian integration and requires the interpolation of values from cell centres to face centres. The interpolation scheme is then given by the *linear* method.

- **Divergence schemes**: As it can be seen in the governing equations, there are two types of terms with divergence operators: one with advection, and the other without advection which have a diffusive nature. The former is discretized by a divergence scheme, while the latter is discretized by a Laplacian scheme. The treatment of the advective terms is one of the major challenges in CFD numerics and so the options are more extensive. For the advetive terms in the left hand side of the volume fraction, continuity, and momentum equations, the upwind scheme (first order) was first used but the results were not appropriate. Then the linear scheme (second order) was tested, and the results were improved a bit but still had some oscillations. Finally the following schemes were chosen which led to the stable results with good convergence: For the first term in the left hand side of the volume fraction equation, the *vanLeer01* scheme (second order) is used, while for the third term the *interfaceCompression* scheme is used. For the advective term of the momentum and continuity equations the *limitedLinearV 1* scheme is utilized. It is also worth mentioning that the scheme *limitedLinear* is more limiting than the *vanLeer*, but has lower accuracy. The term "01" refers to limiting between 0 and 1, but the single number "1" means full limiter compared to the number "0" which means no limiter. For the advective terms in the right hand side of the momentum equation and in the left hand side of energy equation the *linear* scheme is used for all cases.

- **Surface normal gradient**: A surface normal gradient is evaluated at a cell face; it is the component, normal to the face, of the gradient of values at the centres of the two cells that the face connects. The basis of the gradient calculation at a face is to subtract the value at the cell centre on one side of the face from the value in the centre on the other side and divide by the distance. The calculation is second-order accurate for the gradient normal to the face if the vector connecting the cell centres is orthogonal to the face. In this book the *corrected* scheme which guarantees the second-order accuracy by imposing an explicit non-orthogonal correction to the orthogonal component is applied where it is needed.

- **Laplacian schemes**: The *Gauss* method is the only available scheme in OpenFOAM for the discretization terms with Laplacian operator. Therefore the *Gauss linear corrected* which is a second order scheme is used for the simulations.

- **Interpolation schemes**: There are different interpolation schemes in the solver, but it is generally recommended that the *linear* interpolation is the most accurate one for most of the cases, therefore this scheme is used for all simulation cases.

4.4.6   Simulation parameters in *fvSoltion* and *controlDict*

In order to obtain efficiently accurate solutions of the above-mentioned highly coupled set of governing equations, proper solvers and controlling parameters must be selected for each unknown variable. The goal of each solver in OpenFOAM is to solve an equation like AX=B for X in an iterative way with the desired tolerance. To do so in an enhanced and optimized way "preconditioners" are defined which create a Jacobian matrix e.g. M in such way that meets $M^{-1}AX=M^{-1}B$. The matrix M should be invertible easily and cheaply. The best choice is $M=A^{-1}$, but only when A is not costly invertible. In this work the DILU (Diagonal Incomplete LU) preconditioner for non-symmetric matrices is used. The L and U matrices are the lower unitriangular and the upper triangular such that A=LU. In the DILU preconditioner M=LU and A=LU+R, where the matrix R includes all the ignored entries. Two different solvers are used for the simulation of this study. The BiCGStab (modified stabilized biconjugate gradient) solver is used for volume fraction, velocity, and temperature, whereas GAMG the (Generalised geometric-algebraic multi-grid) solver is used for pressure. The selected solvers which are specified in the *fvSolution* dictionary in OpenFOAM are reported in Table 4.2. This selection was obtained after examining various options available in the OpenFOAM software.

Table 4.2: Final settings for simulations.

Variable	Solver	Preconditioner
Volume fraction	BiCGStab	DILU
Pressure	GAMG	-
Velocity	BiCGStab	DILU
Temperature	BiCGStab	DILU

The convergence criteria imposed on the (final) residuals at each physical time step when solving for α, p, $\vec{U}$ and T is 10^{-12}, 10^{-8}, 10^{-8} and 10^{-10}, respectively. These criteria are also set in *fvSoltion*. Furthermore, to achieve the stable solution during solving this highly-coupled set of the governing equations adaptive time stepping is applied with a maximum CFL number set to 0.2, which leads to a maximum time step size of 10^{-5}s. The CFL number as well as the adaptive time step method are set in *controlDict*.

4.5 Solution algorithm

Figure 4.7 shows the flowchart of the algorithm used for the simulations.

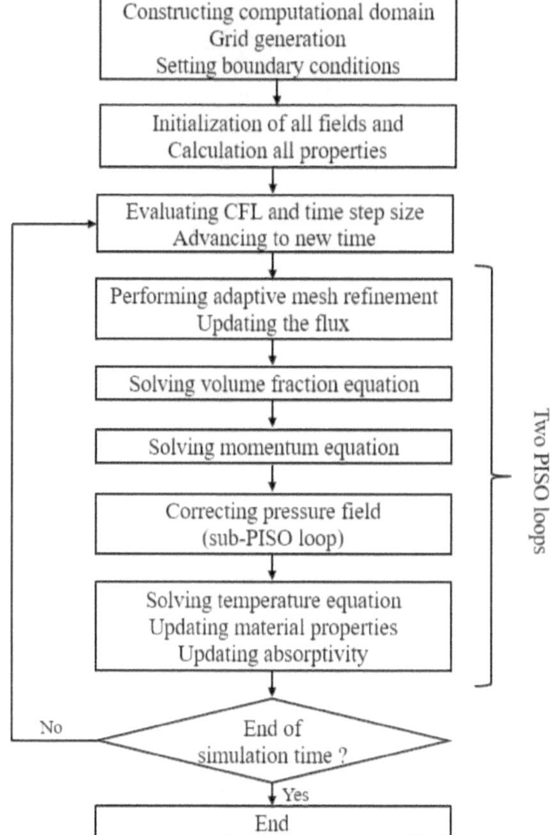

Figure 4.7: Solution procedure.

First the computational domain and the grids are constructed, and then the relevant boundary conditions are set. Based on the initial values all the required parameters are calculated. Then the CFL number and the corresponding time step size (except in the first time step which is determined through an initial value) are calculated. Next the solver goes into a PISO loop with two iterations. In the PISO loop, first adaptive mesh refinement method is applied (to decrease the computational time and yet maintain the accuracy). In this algorithm the interface grids are divided into smaller grids. Next the volume fraction and momentum equations are solved. The pressure field is then updated through a sub-PISO loop with four iterations. This helps the convergence of the solution. Finally the energy equation is solved and all temperature-related parameters are updated accordingly. If the simulation reaches to the final time the solver quits the process, otherwise it goes to the third box in Figure 4.7 and calculates the new CFL and time step size.

Chapter 5

Complementary applications of the developed numerical model

The development of a solver for multi-physics applications raises challenges when it comes to testing the model due to its non-linear and highly coupled nature. Therefore, test cases that are not reported in the appended publications were performed to test sub-parts of the model along its development. The first two test cases are well documented in the literature. In the first one, which is not related to LBW, the focus is on testing the model ability to predict the Marangoni flow while free-surface deformation is quasi-inexistent. The second test case, closer to LBW, is an academic test case since the heat source remains stationary with respect to the workpiece. The temperature dependence of the thermodynamic and transport properties is neglected, and the absorptance is assumed uniform and constant. The applications with temperature dependent material properties, moving heat source, and the proposed absorptance model are presented in the appended papers and are not repeated here. Finally, in test case 3 which complements Paper 3 and is not yet included in a publication, the effect of variable absorptivity on the melt pool geometry is discussed shortly.

5.1   Case 1: Marangoni driven flow with phase change

The first test case is a two-dimensional domain filled with argon gas and bismuth at which phase change phenomenon takes place without any external heat source (e.g. laser). The aim of this test case is to evaluate the performance of the new developments for the energy equation as well as the phase change model for simulating a free surface problem in the presence of the Marangoni force and phase change. The schematic of this test case is illustrated in Figure 5.1. The computational domain is a rectangular cavity which spanned 15 mm and 5 mm in the horizontal (x) and vertical (y) directions, respectively. All problems in OpenFOAM should have three dimensions, even if the case is two dimensional. Therefore the size of the computational domain in the z direction is set to 1 mm (with only one cell). The cavity is filled with Bismuth (bottom region) and Argon (top region). A structured hexahedral grid with the size of 140×62×1 is used for the simulation. The initial location of each phase is shown in Figure 5.1.

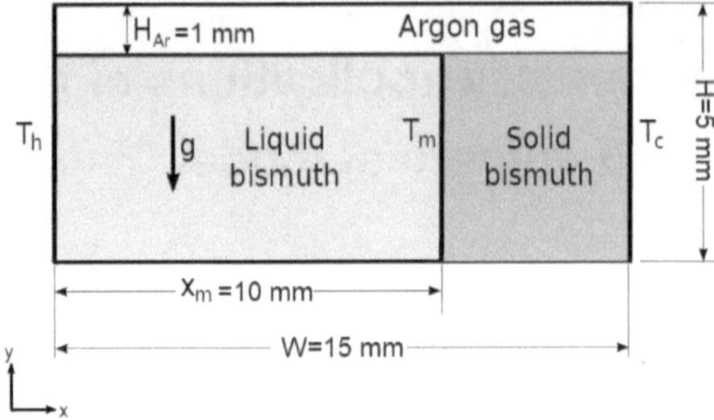

Figure 5.1: Schematic of the validation case 1 [16].

The boundary conditions at the top and bottom walls are:

$$u, v = 0 \tag{5.1}$$

$$\frac{\partial p}{\partial y} = 0 \tag{5.2}$$

$$T = -800x + 552.55 \text{ (linear distribution, } x \text{ is in mm and } T \text{ is in K)} \tag{5.3}$$

$$\frac{\partial \alpha}{\partial y} = 0 \tag{5.4}$$

The boundary conditions at the left and right walls are:

$$u, v = 0 \tag{5.5}$$

$$\frac{\partial p}{\partial x} = 0 \tag{5.6}$$

$$T_{\text{right}} = T_c = 540.55 \text{ K} \tag{5.7}$$

$$T_{\text{left}} = T_h = 552.55 \text{ K} \tag{5.8}$$

$$\frac{\partial \alpha}{\partial x} = 0 \tag{5.9}$$

where u and v are the horizontal and vertical components of velocity vector $\vec{U}$. The dimensionless numbers for this test case are presented in Table5.1. In this case the properties of the working material are also kept constant. Moreover, the melting temperature is set to $T_m = 544.55$ K. The time step size of 10^{-6} s is applied, and the simulation process continues until the residual of each variable reaches to 10^{-8} (steady state condition).

Table 5.1: Dimensionless numbers for the test case 1.

Parameter	Pr	Ma	Ra	Ca	Ste	Bo
Value	0.019	244.1	0.0325	0.0022	0.0331	0.0002

Figure 5.2 compares the present numerical results for the melt front line at steady state condition with the results of Tan et al. [100]. As can be seen, the present solver is well-capable of predicting the location of the liquid-solid interface line.

Figure 5.3 visualizes the temperature and velocity vector fields at the steady state condition. The results show two different vortices, one at the center, and one at the top of the domain. The main vortex which is located in the center of the domain is related to the Marangoni force which arises from the changes in surface tension with respect to temperature on the free surface. As discussed before the direction of this force is from the higher temperature (low surface tension) to the lower temperature (high surface tension). The Marangoni force enhances the heat transfer from the left wall to the colder region and to melt the solid phase, and consequently leads to the main clockwise vortex. The Marangoni force also induces another counter clockwise vortex in the gas phase which spans from the hot wall to the solid-liquid interface.

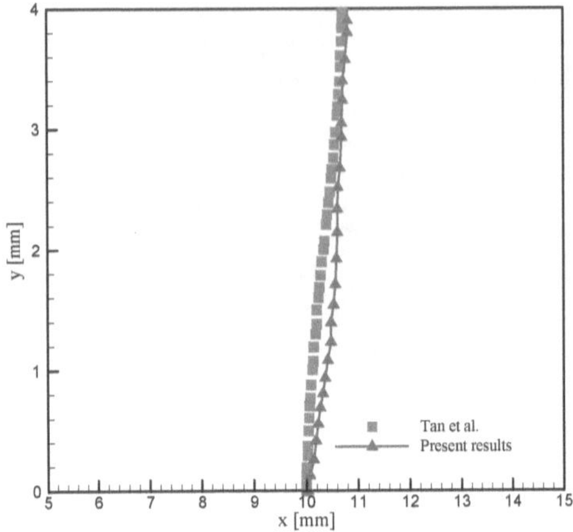

Figure 5.2: Comparison between present numerical results and the results from [100].

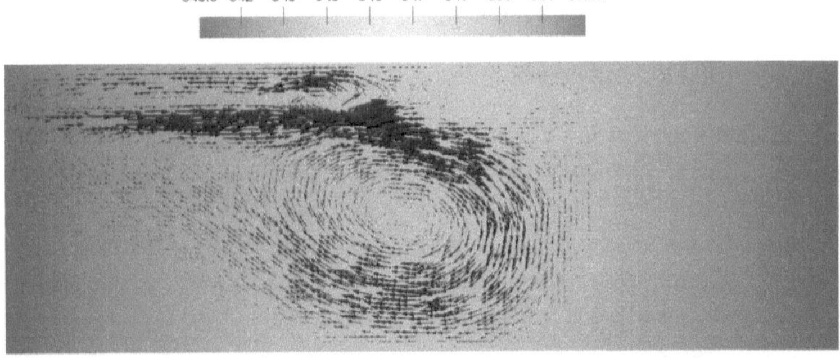

Figure 5.3: Temperature field and velocity vectors for the validation case 1.

The following learning outcomes are reached from this test case:

- Getting familiar with OpenFOAM environment and the solver.

- Assuring that the implementations for the energy equation, surface force, material properties, and phase change model work properly.

5.2 Case 2: Fusion of steel with fixed laser position

In the second test case, which is also a two-dimensional domain now filled with argon gas and a steel alloy, some other new implementations such as laser beam heat source and SSF approach for surface tension related forces are used. Another goal of this test case is to obtain the proper solver settings to reach to a stable solution. In this test case steel is melted with a fixed laser. This case was originally studied both numerically (with neglecting the surface deformation) and experimentally by Pitscheneder et al. [101]. Figure 5.4 shows the schematic of the computational domain. The workpiece is a disk with the size of 15 mm in both x and y directions. The gas phase is located on the top of the disk with the initial thickness of 3.5 mm. Saldi [16] showed that the results for the 3D and axisymmetric cases were almost identical, therefore in this research work only a portion of the disk with 5° sector is considered for simulation. This value is recommended for simulation of axisymmetric geometries in OpenFOAM [88]. A structured hexahedral grid with the resolution of $n_x \times n_y \times n_z = 80 \times 60 \times 1$ in the solid phase and $n_x \times n_y \times n_z = 80 \times 24 \times 1$ in the gas phase is used for the simulation. Saldi [16] also showed that the grids with higher resolutions did not affect the results. As it can be seen in Figure 5.4 that the resolution of the grid is higher at the interface in both solid and/or liquid and gas directions in order to accurately capture the surface deformation and the related phenomena.

The material properties of the steel alloy are presented in Table 5.2 [16]. The top gas area is filled with argon with constant properties (Table 5.3).

Table 5.2: Material properties of the steel alloy used in the test case 2 [16].

Parameter	Value	Dimension
Density	8100	kg/m^3
Dynamic viscosity	0.006	$kg/(m.s)$
Liquid specific heat capacity	723	$J/(kg.K)$
Solid specific heat capacity	627	$J/(kg.K)$
Liquid thermal conductivity	22.9	$W/(m.K)$
Solid thermal conductivity	22.9	$W/(m.K)$
Latent heat of fusion	2.51×10^5	J/kg
Volumetric expansion coefficient	1.5×10^{-6}	$1/K$
Melting temperature	1623	K

The steel alloy of this test case contains surfactants with samples at 20 or 150 ppm sulfur. The model for binary alloy of surface tension with respect to the temperature and sulfur concentration first proposed by Sahoo et al. [102] is therefore applied:

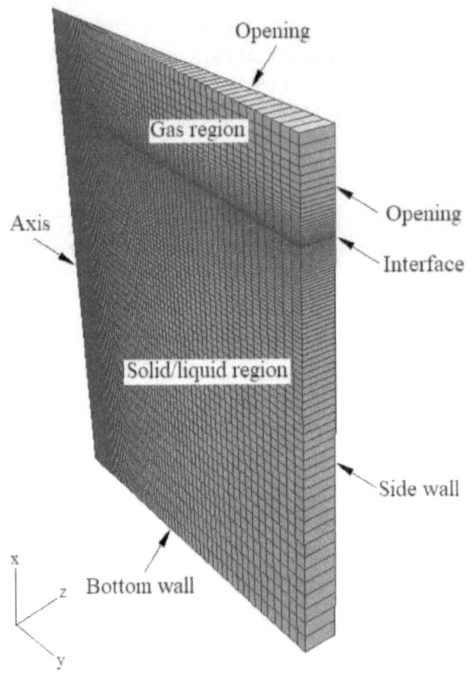

Figure 5.4: Computational domain of the test case 2.

Table 5.3: Material properties of argon used in the test case 2 [16].

Parameter	Value	Dimension
Density	1.6337	kg/m^3
Dynamic viscosity	2.26×10^{-5}	$kg/(m.s)$
Specific heat capacity	520	$J/(kg.K)$
Thermal conductivity	0.0177	$W/(m.K)$

$$\sigma = \sigma_0 + (\frac{d\sigma}{dT})_0(T - T_{melt}) - RT\Gamma_s \ln(1 + Ka_i) \tag{5.10}$$

with

$$\frac{d\sigma}{dT} = (\frac{d\sigma}{dT})_0 - R\Gamma_s ln(1 + Ka_i) - \frac{Ka_i}{1 + Ka_i}\frac{\Gamma_s \Delta H^0}{T} \tag{5.11}$$

$$K = k_1 \exp\left[\frac{-\Delta H^0}{RT}\right] \tag{5.12}$$

where σ_0 and $(\frac{d\sigma}{dT})|_0$ are the surface tension and its variation with respect to temperature at the melting point for pure metal. Furthermore, R is the universal gas constant, Γ_s is the surface excess at saturation, K is the equilibrium constant for segregation, k_1 is the entropy factor, ΔH^0 is the standard heat of adsorption, and a_i is the activity of the element in % weight. Table 5.4 presents the values of these parameters for the Fe-S binary alloy.

Table 5.4: Parameters for calculation of surface tension of Fe-S binary alloy [102].

Parameter	Value	Dimension
$(\frac{d\sigma}{dT})\|_0$	-5×10^{-4}	N/(m.K)
σ_0	1.943	N/m
Γ_s	1.3×10^{-5}	mol/m^2
k_1	0.00318	J/(kg.K)
ΔH^0	-1.66×10^5	J/mol

Figure 5.5 shows the variation of σ and $(\frac{d\sigma}{dT})$ with respect to temperature. The value of $(\frac{d\sigma}{dT})$ changes from positive to negative value at $T = 1735$ K and $T = 2040$ K for the sulfur concentration of 20 ppm and 150 ppm, respectively. It means that the direction of the Marangoni force, and consequently, the flow direction inside the melt pool become reverse.

Figure 5.6 shows the boundary conditions, in the xy plane, for the simulation. The left side is assumed to be the axis at which the gradients of all model variables are zero. The top and top-right surfaces are open to the atmosphere. The bottom and bottom-right surfaces are the boundaries of the workpiece which are assumed to be solid walls.

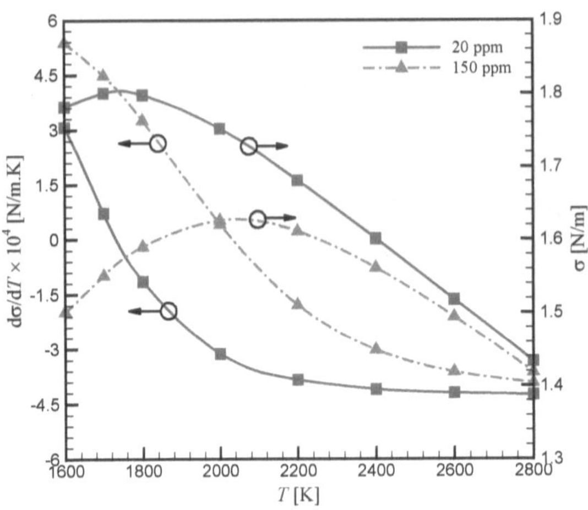

Figure 5.5: Variation of σ and $(\frac{d\sigma}{dT})$ with respect to temperature for the steel alloy at two different concentrations in surfactant.

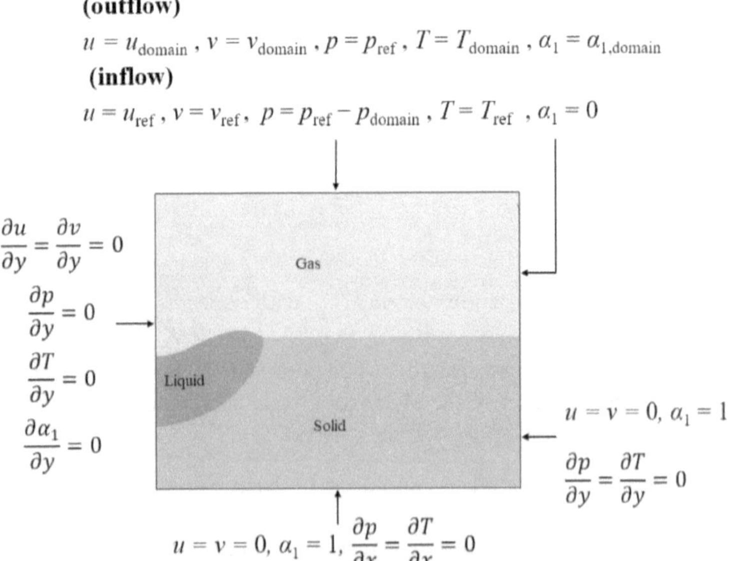

Figure 5.6: Boundary conditions for the test case 2.

Furthermore, the laser heat input is imposed at the gas-liquid/solid interface as:

$$\dot{q}_{\text{laser}} = \begin{cases} \frac{\eta P}{\pi r_{\text{beam}}^2} & if \quad r \leq r_{\text{beam}} \\ 0 & if \quad r > r_{\text{beam}} \end{cases} \tag{5.13}$$

where $\eta = 0.13$ is the laser absorption coefficient, $r_{\text{beam}} = 0.0014$ m is the laser beam radius, and $P = 3850$ W is the laser power. The free surface is cooled by both convection and radiation. The convection effect is automatically considered through the governing equations as there is a direct contact between the top surface of steel and the ambient gas above it. The radiation heat flux is calculated at the free surface as:

$$\dot{q}_{rad} = \epsilon \sigma_B (T^4 - T_{amb}^4) \tag{5.14}$$

where $\epsilon = 0.5$ is the material emissivity, $\sigma_B = 5.67 \times 10^{-8}$ W/(m^{2}K^4) is the Boltzman constant, and $T_{amb} = 300$ K is the ambient temperature. It is also worth mentioning that there was a shielding gas on the top surface of the material in the original experimental work of Pitscheneder et al. [101]. The required information, such as nozzle distance from the surface and its flow direction, was not available so it could not be considered in the present simulations. Therefore, it is expected to observe some deviations in predicting the free surface location, since the shielding gas might impose external shear and normal stresses on the top surface. Besides, Saldi [16] did impose radiative cooling on the top of the computational domain, so at the atmosphere boundary rather than on the material surface. Figure 5.7 shows comparisons between the present numerical results, the numerical results of Saldi [16], and the experimental data of Pitscheneder et al. [101]. It is clear that both numerical methods under-predicted the melt pool depth, but the melt pool width is better predicted in the current numerical simulations. However, the melt pool depth is clearly underestimated. This is a well known problem and the authors use to introduce an enhancement (that can be by a factor of about 10 or more) of the thermal conductivity and sometimes also the viscosity to obtain better agreement between computational and experimental results, To motivate this empirical correction it is presumed that the flow is turbulent. However Kides et al. who investigated the same problem with large eddy simulation (LES) [25] and direct numerical simulation (DNS) [26] concluded that the flow field was transitional. Moreover, including turbulence did not allow them predicting the penetration depth obtained experimentally, nor the melt pool shape. Therefore, an other aspect of the model or test case might need to be questioned.

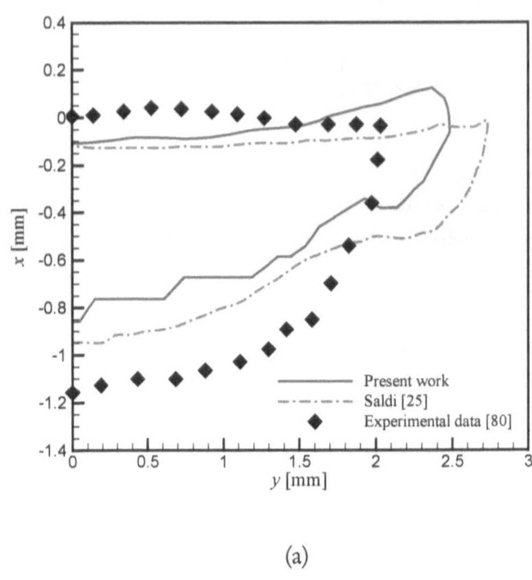

(a)

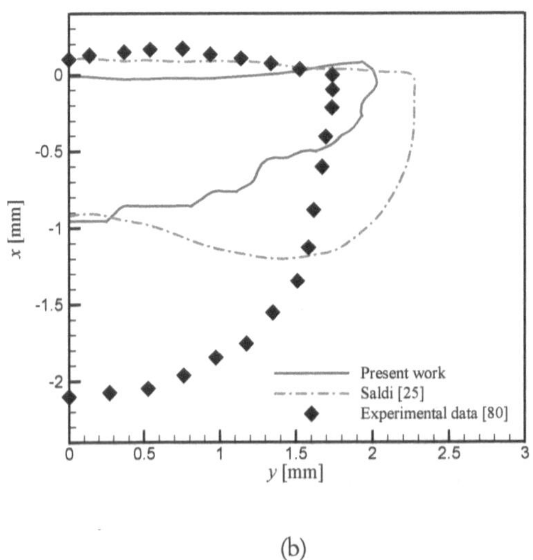

(b)

Figure 5.7: Melt pool profiles at different sulfur concentrations; (a) 20 ppm, (b) 150 ppm.

Although a good agreement was not obtained against the experimental data, the following learning outcomes are reached:

- Implementing a heat source model for laser beam.

- Obtaining the proper solver setting parameters in order to get a stable solution.

- A new test case needs to be designed with a well documented material so as to reduce the uncertainties in the model. Moreover, a very recent study by Ebrahimi et al. [103] did show the importance of considering the temperature dependence of the material properties since it induces variations in the thermocapillary flow. Among the few well-documented metal alloys used in welding, Ti-6Al-4V is selected for conducting this study.

- New validation data are needed, that are qualitative and document also new aspects of the process, including along the welding travel direction. Indeed, the validation of a three-dimensional problem should not be restricted to comparisons with 2-dimensional transverse cross section. This important aspect of the book was conducted in collaboration with the PhD student Yongcui Mi who developed a new measurement method that allowed completing the validation data, as shown in the appended Paper 2.

- As the studies of Kides et al. [25, 25] did show that the discrepancy between computational and calculation results usually attributed to presumed turbulence still exist when accounting for a proper modelling of turbulence, it was concluded in this study that an other part of the model had to be questioned. It was then concluded that the evaluation of the fraction of beam energy absorbed by the material had to be questioned. This did result in the Paper 3.

5.3   Case 3: Effect of non-uniform absorptance

The obtained results in the appended Paper 3 show that considering variable laser absorptance model leads to more accurate results compared to the constant absorptivity coefficient of 0.34 commonly accepted in LBW of Ti-6Al-4V with a 1 μm wavelength laser beam. Moreover, for each of the three test cases investigated in the appended Paper 3 an almost uniform absorptance of 0.38 was obtained. However, it was presumed in the conclusion of Paper 3 that 0.38

should not be considered as an updated reference value of the absorptivity coefficient to be used in all LBW applications with Ti-6Al-4V. The test case of this section is proposed to check whether this presumption can be confirmed.

Here, all the process parameters, boundary conditions, and assumptions are identical to the ones that were used in the appended Papers 2 and 3 except that a laser with a total power of 500 W and an elliptical beam profile with major axis along the welding direction are used. The minor and major radii of the beam profile are 0.658 mm and 0.875 mm, respectively. The laser travel speed is also considered to be 10 mm/s.

Figure 5.8 depicts the top view of the melt pool when it is at stable condition ($t > 0.7$ s). The top picture shows the contours of laser absorption while the bottom one shows the variations of temperature. It also provides some iso-lines of power density (W/m^2). The results show that the laser absorption varies from 0.365 to 0.377 inside the laser spot. So it is seen that A=0.38 is outside the predicted range of absorptance. It is however quite close. Now the question is if a constant value of $A = 0.38$ can lead to the same results with the new beam characteristics?

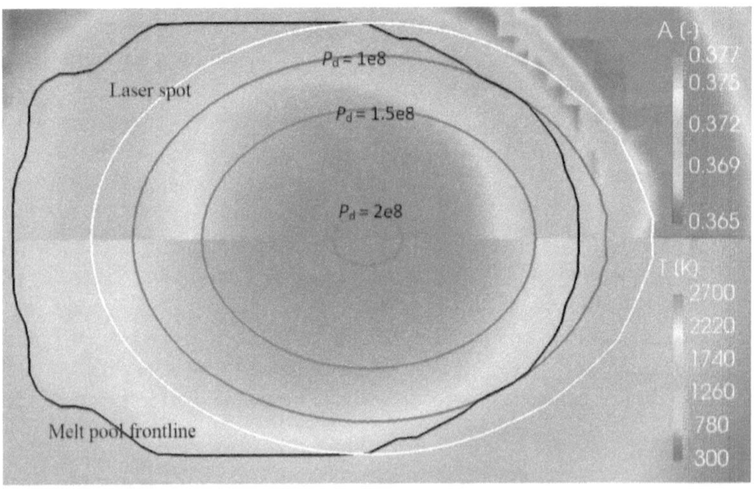

Figure 5.8: Melt pool top-view (welding direction is from left to right, and the presented power density distribution (P_d) is in W/m^2) .

Figures 5.9 and 5.10 can provide useful information in order to give a primary response to the above question. As can be seen in Figure 5.9, the melt pool free surface deformation is slightly higher when using constant absorption (due to higher temperature gradient and so bigger thermal Marangoni force). Another point is that the melt pool width is obviously smaller when the variable absorption approach is utilized.

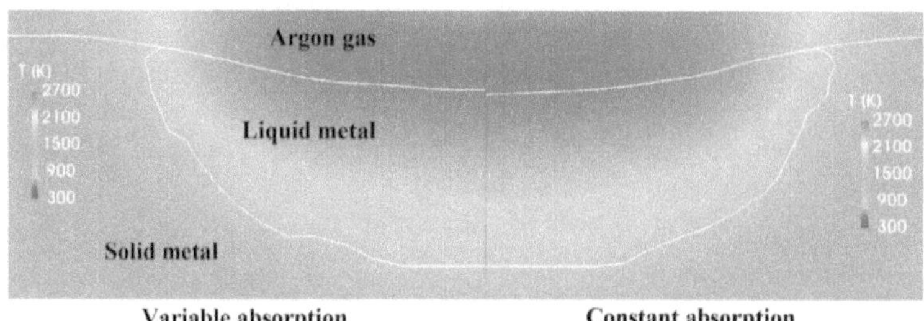

Figure 5.9: Cross-sectional view of the melt pool predicted by using constant absorptivity coefficient of $A = 0.38$ (right) and variable absorptance (left).

Taking a look at the melt pool volume evolution during time (Figure 5.10) shows that the predicted melt pool volume at the quasi-steady condition is about 7% larger when a constant absorption of $A = 0.38$ is used.

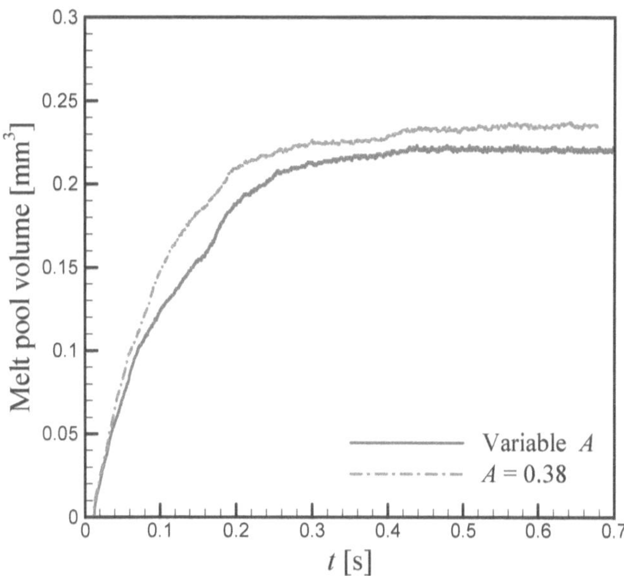

Figure 5.10: Variations of melt pool volume with respect to time.

The presented results partially confirm the claim in the appended Paper 3 about the doubt on the general validity of $A = 0.38$ as an average value of laser absorption for welding of Ti-6Al-4V in conduction mode. This research question needs to be further investigated to establish how the non-uniformity of the absorptivity in the beam spot affects the flow field inside the melt pool. Ongoing simulations will be analysed in the second part of this PhD book in order to provide a complementary response to RQ3.

Chapter 6

Summary of the appended papers

The main results of this research work have been organized into the three following papers intended to be published in international journals.

1. Effect of different shaped laser beam profiles on melt pool dynamics and geometry in conduction welding.
   Status: Published in the International Journal of Thermal Sciences.
   Authors: S. M. A. Noori Rahim Abadi, Y. Mi, F. Sikström, A. Ancona, I. Choquet.
   Author contribution: Seyyed Mohammad Ali Noori Rahim Abadi performed the relevant solver developments and numerical simulations. He also post-processed the results and prepared the first draft of the article.
   Relevant to the research question one
   Summary: Laser beam shaping is a powerful developing approach enabling tailoring fusion welding to improve the weld seam. In this study a computational fluid dynamic approach was used to investigate the influence of beam shaping on the melt thermal flow and geometry. A solver, capable of tracking the deformation of the melt free surface was developed in an open source software. It was tested with bead on plate conduction welds in Ti-6Al-4V with a top-hat beam profile and provides good agreement with the experimental measurements. Three beam shapes were studied at several welding travel speeds: a Gaussian profile and two elliptic elongations of this profile (ratio ≈ 3.8) one along and one transverse to the welding travel direction.
   Main conclusions: It was found that these beam shapes clearly change the melt flow, and the melt flow contributes to changing the melt pool geometry. In particular, the beam profile elongated along the welding travel direction generates melt pool front vortices that assist metal preheating, and can result in penetration depth as deep as with the Gaussian profile for low welding travel speed, while the free surface deformation

and the amount of thermal energy diffused in the HAZ behind the melt pool can be lower depending on the travel speed.

2. Dynamic free surface contour measurement of the melt pool applied to conduction mode laser beam welding of Ti-6Al-4V and model evaluation.
Status: Submitted.
Authors: Y. Mi, S. M. A. Noori Rahim Abadi, F. Sikström, I. Choquet, A. Ancona.
Author contribution: Seyyed Mohammad Ali Noori Rahim Abadi performed the relevant solver developments and numerical simulations. He post-processed the numerical results in order to be compared against the experimental data. He also contributed to prepare the section for numerical simulation in the paper.
Relevant to the research questions two and three
Summary: This work provides new ways to experimentally acquire both in-process and post-process data for model validation of conduction mode LBW. An imaging acquisition setup equipped with narrow bandpass filter and illuminating LEDs was designed, built and calibrated to measure the melt pool geometries with an accuracy of less than 100 μm. The method was applied to LBW of Ti-6Al-4V for different travel speeds. Melt pool shapes, width, length, and area were measured in world coordinates. The speed impact on the melt pool width and length stability was also analysed by the corresponding measurement distribution. The solidification time of the ending melt pool and the morphology of the ending crater were acquired for model validation as well. All these characterizations provide the possibility to evaluate a model from multiple aspects.
Main conclusions: This work was initially performed to evaluate a numerical model investigating the effects of different welding speeds and laser absorption coefficients on the melt pool size. It could also be used as a reference approach to evaluate other numerical simulation models in a more complete way.

3. Modelling of beam energy absorbed locally in conduction mode laser metal fusion.
Status: Submitted.
Authors: S. M. A. Noori Rahim Abadi, Y. Mi, F. Sikström, I. Choquet.
Author contribution: Seyyed Mohammad Ali Noori Rahim Abadi performed the relevant solver developments and numerical simulations. He also post-processed the results and prepared the first draft of the paper.
Relevant to the research questions two and three
Summary: In this work a method is proposed in order to calculate the laser absorptance with respect to the laser wavelength, surface tempera-

ture and curvature. The Maxwell's equations are used to define the laser electromagnetic field, and then are simplified with a set of assumptions. The proposed method is used in the LBW of Ti-6Al-4V. To do so the absorptivity calculation procedure as well as the melting and solidification process due to laser heat source are implemented into a new solver in OpenFOAM software. The effect of variable laser absorptivity at different welding speeds is investigated, and the results are compared against the experimental data and the ones when the absorptivity is simplified to a constant value.

Main conclusions: The results showed that the constant absorptivity coefficient of 0.34 commonly accepted in LBW of Ti-6Al-4V with a 1 μm wavelength laser beam under-predicted the melt pool size. Furthermore, it was found that applying variable laser absorptance led to predicting the more accurate melt pool size compared to the experimental measurements.

Chapter 7

Conclusion

The main purpose of this research work was to better understand the physics of melt pool thermal-hydrodynamics behaviour during leaser beam welding. The gained knowledge would be used to e.g. process control. To do so, a new solver in OpenFOAM software was developed in order to simulate the laser beam welding of Ti-6Al-4V in conduction mode without metal transfer. The work had been divided into several tasks:

- Firstly, the temperature-dependent material properties of Ti-6Al-4V were implemented into the solver. In order to find reliable database for each material property, numerous published works have been checked. The most commonly used values were chosen for each property.

- Secondly, different non-stationary laser beam heat sources were added to the solver. In total four continuous (one Gaussian, one top-hat, and two elliptical) and one pulsing laser beam profiles were implemented into the code.

- Thirdly, a new calculation process was implemented into the solver to predict the laser beam absorptance with respect to the local temperature, surface curvature, and laser beam wavelength.

- Finally, various test cases were simulated in order to first evaluate the accuracy of the new developed solver, and then investigate the effect of variable laser absorptivity and laser beam profile on the melt pool thermo-hydrodynamic behaviour.

7.1 Main findings

The obtained results can be used to answer the three main research questions of this research work as:

- **RQ1: How do different power density distributions in the laser beam spot modify the process physics during laser beam welding in conduction mode?**

  The simulation results confirmed previous known knowledge as:

  (1) For each studied beam profile an increase in the welding travel speed decreases the melt pool width and penetration depth.

  (2) The Gaussian beam profile, with significantly higher peak intensity, leads to the highest melt pool temperature.

  (3) The transverse Gaussian-elliptic profile leads to the widest, and shallowest melt pool.

  The new findings can be summarized as:

  (1) The three beam shapes studied induce melt flow patterns with clear specificities, implying that in these cases beam shaping changes not only heat conduction but also heat convection (RQ1.1).

  (2) The Gaussian profile leads to wider melt pool than its longitudinal elliptic elongation for the large ellipse-aspect ratio of this study (RQ1.2).

  (3) While the transverse and longitudinal Gaussian-elliptic beam profiles can be obtained from each other by rotation, their melt pools (both flow pattern and geometry) cannot. Moreover, their melt pool volumes differ significantly. Melt convection has thus a leading order role when studying the effect of beam shaping in conduction mode laser beam welding (RQ1.2 and RQ1.3).

  (4) Among the beam profiles investigated in this study, the profile elongated along the welding travel direction is the only one to induce an additional vortex at each side (by symmetry) of the melt pool front. This vortex assists the fusion process through pre-heating the alloy, and enhances both penetration depth and fused volume. As a counterpart it lowers the amount of thermal energy dissipated in the heat affected zone. These effects are more pronounced at low welding travel speed at which they can compete with or even outdo the large penetration depth characteristic of the Gaussian profile, while the free surface deforms less with the elongated profile. On the contrary, for the Gaussian and transverse elliptic beam shapes no change in flow pattern was observed when lowering the power density (RQ1.3).

- **RQ2: How can the absorptivity of laser beam energy by a metal alloy be predicted when simulating laser beam welding in conduction mode with CFD?**

It is recalled that this research question was formulated based on the results presented in Paper 2. The absorptivity coefficient of 0.34 commonly accepted in conduction mode laser beam welding of Ti-6Al-4V with a Nd:YAG laser result in under-predicting the melt pool size (Paper 3). This is the reason why a larger absorptivity coefficient of 0.4 was selected (after trial and error search) to try to best fit the experimental measurements in Paper 2. But a trial and error approach to fit experimental measurements implies that the simulation model is time consuming, limited by the availability of experimental data, and more importantly not predictive. Therefore, another approach was investigated (RQ2.1).

Going back to the physical principles governing the absorption of electromagnetic energy by a material, and imposing the constraint of modelling at a thermo-fluid scale (not at a microscopic scale), the absorptance was modelled as a function of local surface properties that are its curvature and temperature (Paper 3). It is recalled that the surface curvature defines the local angle of incidence of the beam (RQ2.2).

It was obtained (Paper 3) that for Nd:YAG laser and Ti-6Al-4V material the dependence of the absorptance with respect to the local beam incidence angle is confirmed to be negligible since surface deformation is weak, implying that the range of incidence angle is $\theta_i \leq 30°$. Therefore, the angular dependence of the absorptance can be neglected in conduction mode laser beam welding. However, it should not be neglected if the conduction mode laser beam welding is performed depositing metal with a wire since then the range of incidence angle extends beyond 30°. For Nd:YAG laser and Ti-6Al-4V the absorptance shows a clear temperature dependence over a wide range of incidence angle, from normal incidence to almost grazing incidence. Therefore, the temperature dependence of the absorptance should be taken into account when modelling conduction mode laser beam welding with a CFD approach. This conclusion is believed to apply also when the laser beam fusion process is supplemented with a wire to deposit metal (RQ2.3).

- **RQ3: How does a predicted absorptance affect the process physics during LBW compared to a constant absorptivity coefficient?**

It could be seen that the melt pool geometry is very sensitive to the absorptance value. It was also observed that for the test cases with uniform power density distribution of Paper 3 the absorptance model predicts a laser beam absorptivity that is almost uniform in the beam spot. This value of 0.38 was larger than the absorptivity coefficient of 0.34 that is commonly accepted in conduction mode laser beam welding of Ti-6Al-4V with a Nd:YAG laser. It was also seen in Section 5.3 that different process conditions can lead to different ranges of variation of the absorptance in the beam spot, and to different mean absorptance in the beam spot (RQ3.1).

The predicted non-uniform value was used to approximate an equivalent uniform absorptivity coefficient. It could be seen that melt pool CFD models computed with both non-uniform absorptance and equivalent uniform absorptivity coefficient lead to some differences in the melt pool geometry.

It is known that temperature variations on the melt pool surface change the thermocapillary flow. And in conduction mode laser beam welding the thermocapillary acceleration is a leading order source of convection. Therefore, it is believed that laser beam welding with process conditions leading to a wider absorptivity range within the beam spot might show a larger influence of the absorptivity variation within the beam spot on the melt pool. This needs to be further investigated, for instance from the test case 3 presented in the previous chapter. Based on the computational results obtained with the variable absorptance model, an equivalent absorptivity coefficient that is representative of the predicted absorptivity range (such as a mid-range value) should be determined. Then, the computational results obtained with variable absorptance and equivalent uniform absorptivity coefficient should be analysed to complete the answer to this research question (RQ3.2).

7.2 Future works

In the continuation of this PhD book, more simulations with different process conditions are first needed to further investigate the effect of variable absorption, related to RQ3. Then, the new knowledge as well as the new implemented items to the solver will be used to simulate the metal fusion with the presence of a wire to extend the model to metal transfer. This will be in the context of metal AM. The laser heat source will be used to fuse both substrate and wire. This procedure will then be repeated for specific number of deposited layers to build a small part. The effect of various parameters such as wire feeding rate and angle, wire pre-heating, laser beam angle and travel speed, and potential role of laser beam reflection on the melt flow, melt pool geometry, and bead profile will be investigated.